Probability and Mathematical Statistics (Continued)

SCHEFFE · The Analysis of Variance
SEBER · Linear Regression Analysis
WILKS · Mathematical Statistics
WILLIAMS · Diffusions, Markov Processes, and Martingales,
Volume I: Foundations
ZACKS · The Theory of Statistical Inference

Applied Probability and Statistics

BAILEY · ochastic Processes with Applications
to t' a
F. L. M. istics and Systems for Health
l and LEWIS · Outliers in Statistical Data
b MFW · Stochastic Models for Social Processes, *Second*

BARTHOLOMEW and FORBES · Statistical Techniques in Man-
power Planning
BECK and ARNOLD · Parameter Estimation in Engineering and
Science
BELSLEY, KUH, and WELSCH · Regression Diagnostics: Identi-
fying Influential Data and Sources of Collinearity
BENNETT and FRANKLIN · Statistical Analysis in Chemistry and
the Chemical Industry
BHAT · Elements of Applied Stochastic Processes
BLOOMFIELD · Fourier Analysis of Time Series: An Introduction
BOX · R. A. Fisher, The Life of a Scientist
BOX and DRAPER · Evolutionary Operation: A Statistical Method
for Process Improvement
BOX, HUNTER, and HUNTER · Statistics for Experimenters: An
Introduction to Design, Data Analysis, and Model Building
BROWN and HOLLANDER · Statistics: A Biomedical Introduction
BROWNLEE · Statistical Theory and Methodology in Science and
Engineering, *Second Edition*
BURY · Statistical Models in Applied Science
CHAMBERS · Computational Methods for Data Analysis
CHATTERJEE and PRICE · Regression Analysis by Example
CHERNOFF and MOSES · Elementary Decision Theory
CHOW · Analysis and Control of Dynamic Economic Systems
CLELLAND, deCANI, BROWN, BURSK, and MURRAY · Basic
Statistics with Business Applications, *Second Edition*
COCHRAN · Sampling Techniques, *Third Edition*
COCHRAN and COX · Experimental Designs, *Second Edition*
COX · Planning of Experiments
COX and MILLER · The Theory of Stochastic Processes, *Second
Edition*
DANIEL · Biostatistics: A Foundation for Analysis in the Health
Sciences, *Second Edition*
DANIEL · Applications of Statistics to Industrial Experimentation
DANIEL and Wood · Fitting Equations to Data: Computer
Analysis of Multifactor Data, *Second Edition*
DAVID · Order Statistics
DEMING · Sample Design in Business Research
DODGE and ROMIG · Sampling Inspection Tables, *Second Edition*
DRAPER and SMITH · Applied Regression Analysis

continued on back

Regression Analysis by Example

To Allegra, Ann, Anne, Martha, and Rima

Preface

Regression analysis has become one of the most widely used statistical tools for analyzing multifactor data. It is appealing because it provides a conceptually simple method for investigating functional relationships among variables. The standard approach in regression analysis is to use a sample of data to compute an estimate of the proposed relationship, and then evaluate the fit using statistics such as t, F, and R^2.

Our approach is much broader. We view regression analysis as a set of data analytic techniques that are used to help understand the interrelationships among a given set of variables. The emphasis is not on formal statistical tests and probability computations. We argue for an informal analysis directed towards uncovering patterns in the data.

We utilize most standard and some not so standard summary statistics on the basis of their intuitive appeal. We are not overly concerned with precise probability evaluations. We rely heavily on graphical representations of the data. In particular, many variations of plots of regression residuals are used. Graphical methods for exploring residuals can suggest model deficiencies or point to "troublesome" observations. Upon further investigation into their origin, the troublesome observations often turn out to be more informative than the well-behaved observations. We feel that more information is obtained from an informal examination of a plot of residuals than from the formal test of statistical significance of some limited null-hypothesis. In short, our presentation in the chapters of this book is guided by the principles and concepts of exploratory data analysis.

The various concepts and techniques of regression analysis are developed with the aid of examples. In each example, we have isolated one or two techniques and discussed them in some detail. The data was chosen to highlight the techniques being presented. Although when analyzing a given set of data it is usually necessary to employ many techniques, we have tried to choose the various data sets so that it would not be necessary to discuss the same technique more than once. Our hope is that after working through the book, the reader will be ready and able to analyze his or her own data methodically, thoroughly, and confidently.

No attempt is made to derive the techniques used. Techniques are described, the required assumptions are given, and finally, the success of the technique in the particular example is assessed. Although derivations of the techniques are not included, we have tried to refer the reader in each case to sources in which such discussion is available. Our hope is that some of these sources will be investigated by the reader who wants a more thorough grounding in theory. The emphasis in this book is not on formulas, tests of hypotheses, or confidence intervals, but on the analysis of data.

We have taken for granted the availability of a computer and a statistical analysis system. We feel that there has been a qualitative change in the analysis of linear models, from model fitting to model building, from overall tests to clinical examinations of data, from macroscopic to the microscopic analysis. To do this kind of analysis a computer is essential and we have assumed its availability. No specific machines or plotters are needed to carry out the different analyses. Almost all of the analyses we use are now available in software packages at most computer centers in universities, business, and government agencies.

The material presented is intended for anyone who is involved in analyzing data. The book should be helpful to anyone who has some knowledge of the basic concepts of statistics. In the university, it could be a text for a course in regression analysis for students whose specialization is not statistics, but nevertheless use regression analysis quite extensively in their work. For students whose major emphasis is statistics, and who take a course on regression analysis from a book at the level of Searle, Plackett, or Rao, this book can be used to balance and complement the theoretical aspects of the subject with practical applications. Outside the university, this book can be profitably used by those people whose present approach to analyzing multifactor data consists of looking at standard computer output (t, F, R^2, standard errors, etc.), but who want to go beyond these summaries for a more thorough analysis.

We have attempted to write a book for a group of readers with diverse backgrounds. We have also tried to put emphasis on the art of data analysis rather than on the development of statistical theory. We are fortunate to have had assistance and encouragement from several friends, colleagues, and associates. We particularly want to mention Professor Martin J. Gardner of the University of Southampton, U. K. who read an early draft and made valuable comments. Some of our colleagues at New York University have used portions of the material in their courses and have shared with us their comments and comments of their students. The students in our classes on regression analysis have all contributed by asking penetrating questions and demanding meaningful and understandable answers. Mr. Tak Lo has assisted with some aspects of the computer work. Ms. Roberta Mollot typed the final manuscript with haste and accuracy. To each of these people and to the many others that have provided assitance and encouragement we are grateful and express our thanks.

We are grateful to the Literary Executor of the late Sir Ronald A. Fisher, F.R.S., to Dr. Frank Yates, F.R.S., and to Longman Group Ltd., London, for permission to reprint four columns from Table III (p. 46) from their book, *Statistical Tables for Biological, Agricultural, and Medical Research* (6th edition, 1974).

SAMPRIT CHATTERJEE
BERTRAM PRICE

Eagle Island, Maine
White Plains, New York
March 1977

Contents

CHAPTER 1   Simple Linear Regression,   1

    1.1   Introduction,   1
    1.2   Description of the Data and Model,   2
    1.3   Estimation and Tests of Hypotheses,   3
    1.4   Index of Fit,   6
    1.5   Predicted Values and Standard Errors,   7
    1.6   Evaluating the Fit,   7
    1.7   Analysis of Residuals,   9
    1.8   Repair Times for Computers,   10
         Bibliographic Notes,   18
         References,   18

CHAPTER 2   Detection and Correction of Model Violations: Simple
Linear Regression,   19

    2.1   Introduction,   19
    2.2   Effects of Outliers in Simple Regression,   19
    2.3   Television Rating Data,   20
    2.4   Model Adequacy and Residual Plots,   23
    2.5   Deletion of Data Points,   25
    2.6   Transformation of Variables,   27
    2.7   Transformations to Achieve Linearity,   29
    2.8   Bacteria Deaths due to X-ray Radiation,   32
    2.9   Transformations to Stabilize Variance,   38
    2.10 Injury Incidents in Airlines,   40
    2.11 An Industrial Example,   44
    2.12 Removal of Heteroscedasticity,   47
    2.13 Principle of Weighted Least Squares,   49
    2.14 Summary,   50
         References,   50

CHAPTER 3   Multiple Regression Model,   51

3.1   Description of the Data and Model,   51
3.2   Properties of the Least Square Estimators,   53
3.3   Predicted Values and Standard Errors,   54
3.4   Multiple Correlation Coefficient,   55
3.5   Tests of Hypotheses in the Linear Model,   56
3.6   Assumptions About the Explanatory Variables,   58
3.7   A Study of Supervisor Performance,   59
3.8   Testing a Subset of Regression Coefficients Equal to Zero,   65
3.9   Testing the Equality of Regression Coefficients,   66
3.10   Estimating and Testing of Regression Parameters under Constraints,   68
3.11   Summary,   70
       References,   70
       Appendix,   71

CHAPTER 4   Qualitative Variables as Regressors,   74

4.1   Introduction: Indicator Variables,   74
4.2   Salary Survey Data,   75
4.3   Systems of Regression Equations: Comparing Two Groups,   85
4.4   Dummy Variables: Other Applications,   95
4.5   Seasonality,   95
4.6   Stability of Regression Parameters over Time,   96
       References,   100

CHAPTER 5   Weighted Least Squares,   101

5.1   Introduction,   101
5.2   Heteroscedastic Models,   102
5.3   Supervisor Data,   102
5.4   College Expense Data,   103
5.5   Two-Stage Estimation,   105
5.6   Education Expenditure Data,   107
5.7   Fitting a Dose–Response Relationship Curve,   115
5.8   The Logistic Model,   117
5.9   Fitting a Logistic Response Function,   118
5.10   Toxicity of Rotenone,   120
       References,   122

CHAPTER 6   The Problem of Correlated Errors,   123

            6.1   Introduction: Autocorrelation,   123
            6.2   Consumer Expenditure and Money Stock,   124
            6.3   Durbin–Watson Statistic,   125
            6.4   Removal of Autocorrelation by Transformation,
                   128
            6.5   Iterative Estimation with Autocorrelated Errors,
                   129
            6.6   Autocorrelation and Missing Variables,   131
            6.7   Analysis of Housing Starts Data,   132
            6.8   Limitations of Durbin–Watson Statistic: Ski
                   Equipment Sales,   136
            6.9   Examining Residual Plots,   137
            6.10   Dummy Variables to Remove Seasonality,   139
                   References,   142

CHAPTER 7   Analysis of Collinear Data,   143

            7.1   Introduction,   143
            7.2   Effects on Inference,   144
            7.3   Effects on Forecasting,   151
            7.4   Detection of Multicollinearity,   155
            7.5   Principal Components in Detection of
                   Multicollinearity,   157
            7.6   Correction for Multicollinearity: Imposing
                   Constraints,   163
            7.7   Searching for Linear Function of the β's,   166
            7.8   The Principal Components Approach,   167
            7.9   Computations Associated with Principal
                   Components,   170
                   Bibliographic Notes,   172
                   References,   172
                   Appendix,   172

CHAPTER 8   Biased Estimation of Regression Coefficients,   175

            8.1   Introduction,   175
            8.2   Principal Components Regression,   176
            8.3   Removing Dependence among the Explanatory
                   Variables,   177
            8.4   Constraints on the Regression Coefficients,   180
            8.5   Ridge Regression,   181

8.6   Definition and Computation,   181
8.7   Detection of Multicollinearity using Ridge Methods, 182
8.8   Estimation by the Ridge Method,   185
8.9   Summary,   187
       Bibliographic Notes,   188
       References,   188
       Appendix,   188

CHAPTER 9   Selection of Variables in a Regression Equation,   193

9.1   Introduction,   193
9.2   Formulation of the Problem,   194
9.3   Consequences of Deletion of Variables,   194
9.4   Preliminary Remarks on Variable Selection,   196
9.5   Uses of Regression Equations,   196
9.6   Criteria for Evaluating Equations,   197
9.7   Residual Mean Square,   197
9.8 C_p: Definition and Use,   198
9.9   Examination of Collinearity,   199
9.10   Evaluating All Possible Equations,   200
9.11   Selection of Variables: Stepwise Procedure,   201
9.12   Forward Selection Procedure,   201
9.13   Backward Elimination Procedure,   201
9.14   Stepwise Method,   202
9.15   General Comments on Stepwise Procedures,   202
9.16   A Study of Supervisor Performance,   203
9.17   Variable Selection with Collinear Data,   206
9.18   Application of Ridge Regression to Variable Selection,   208
9.19   Selection of Variables in an Air Pollution Study, 209
       Bibliographic Notes,   214
       References,   214
       Appendix,   215

STATISTICAL TABLES,   218

INDEX,   225

Regression Analysis by Example

CHAPTER 1

Simple Linear Regression

1.1. INTRODUCTION

Regression analysis may be broadly defined as the analysis of relationships among variables. It is one of the most widely used statistical tools because it provides a simple method for establishing a functional relationship among variables. The relationship is expressed in the form of an equation connecting the response or dependent variable y, and one or more independent variables, $x_1, x_2, \ldots, x_p$. The equation, or to be more precise, the regression equation takes the form

$$y = b_0 + b_1 x_1 + b_2 x_2 + \cdots + b_p x_p,$$

where $b_0, b_1, b_2, \ldots, b_p$ are called the regression coefficients and are determined from the data. A regression equation containing only one independent variable is called a simple regression equation. An equation containing more than one independent variable is referred to as a multiple regression equation. An example of simple regression would be an analysis in which the time to repair a machine is studied in relation to the number of components to be repaired. Here we have one dependent variable (time to repair the machine) and one independent variable (number of components to be repaired). An example of a very complex multiple regression situation would be an attempt to explain the age-adjusted mortality rates prevailing in different geographic regions (dependent variable) by a large number of environmental and socioeconomic factors (independent variables). Both types of problems are treated in the text. In fact, these two particular examples are included, one in the first chapter, the other in the last chapter.

The explicit determination of the regression equation is in a sense the

1

final product of the analysis. It is a summary of the relationship between y (the dependent variable) and the set of independent variables, the x's. The equation may be used for several purposes. It may be used to evaluate the importance of individual x's, to analyze the effects of policy that involves changing values of the x's, or to forecast values of y for a given set of x's. Although the regression equation is the final product, there are many important by-products. We view regression analysis as a set of data analytic techniques that are used to help understand the interrelationships among variables in a certain environment. It is assumed that data from the environment is available. Sometimes the data will have been collected in a controlled setting so that factors that are not of primary interest can be held constant. Most often the data will have been collected under nonexperimental conditions where very little can be controlled by the investigator. The task of regression analysis is to learn as much as possible about the environment represented by the data. We emphasize that what is uncovered along the way to the formulation of the equation may often be as valuable and informative as the final equation.

We begin our study by considering the simple linear regression model. In this chapter the model is formulated, assumptions are stated, and the standard theoretical results are recorded. There are no formal derivations. Familiarity with standard results is developed through the examples. Formulas are presented, but only for purposes of reference. It is assumed throughout that the necessary summary statistics will be computer generated from an existing regression package.* The reader familiar with the basic concepts of regression analysis may choose to begin with the section labeled, "Analysis of Residuals" and then proceed to the example on p. 10, referring back to the formulas as necessary. Readers interested in mathematical derivations are referred to the bibliographic notes at the end of this chapter where a number of books that contain a formal development of the regression problem are listed.

1.2. DESCRIPTION OF THE DATA AND MODEL

The data consists of n observations on a dependent or response variable y and an independent or explanatory variable x_1. The observations are

---

*Most computer centers and commercial computer time vendors offer one or more regression analysis packages. We assume that these programs have been thoroughly tested and produce numerically accurate answers. For the most part the assumption is a safe one, but for some data sets, different programs have given dramatically different results. See Beaton, Rubin, and Barone (1976), or Longley (1967), for a discussion of this problem.

usually recorded as follows:

Observation number	y	x_1
1	y_1	x_{11}
2	y_2	x_{12}
3	y_3	x_{13}
.	.	.
.	.	.
.	.	.
n	y_n	x_{1n}

The relationship between y and x_1 is formulated as a linear* mòdel

$$y_i = \beta_0 + \beta_1 x_{1i} + u_i, \qquad i = 1, 2, \ldots, n, \tag{1.1}$$

where β_0 and β_1 are constants and are called the model regression parameters, and u_i is a random disturbance. It is assumed that in the range of the observations studied, the linear equation (1.1) provides an acceptable approximation to the true relation between y and x_1. In other words, y is approximately a linear function of x_1, and u measures the discrepancy in that approximation. It is assumed that for every fixed value of x_1, the u's are random quantities independently distributed with mean zero and a common variance denoted by σ^2. The coefficient β_1 may be interpreted as the increment in y corresponding to a unit increase in x_1.

1.3. ESTIMATION AND TESTS OF HYPOTHESES

The parameters β_0 and β_1 are estimated by the method of least squares which involves minimizing the sum of squares of the residuals $S(\beta_0, \beta_1)$, where

$$S(\beta_0, \beta_1) = \sum_{i=1}^{n} u_i^2 = \sum_{i=1}^{n} (y_i - \beta_0 - \beta_1 x_{1i})^2.$$

The values of β_0 and β_1 that minimize $S(\beta_0, \beta_1)$, b_0, and b_1 are given by

$$b_1 = \frac{\sum (y_i - \bar{y})(x_{1i} - \bar{x}_1)}{\sum (x_{1i} - \bar{x}_1)^2} \tag{1.2}$$

---

*The adjective linear has a dual role here. It may be taken to describe the fact that the relationship between y and x_1 is linear. More generally, the word linear refers to the fact that the regression parameters enter Equation (1.1) in a linear fashion. As we shall encounter later, $y = \beta_0 + \beta_1 x_1^2 + u$ is also a linear model even though the relationship between y and x_1 is quadratic.

and

$$b_0 = \bar{y} - b_1 \bar{x}_1, \tag{1.3}$$

where

$$\bar{y} = \frac{\sum y_i}{n} \quad \text{and} \quad \bar{x}_1 = \frac{\sum x_{1i}}{n}.$$

Based on the assumptions described previously concerning the u's, it follows that the quantities, b_0 and b_1, are unbiased estimates of β_0 and β_1. Their variances are

$$\text{Var}(b_1) = \frac{\sigma^2}{\sum (x_{1i} - \bar{x}_1)^2}, \tag{1.4}$$

$$\text{Var}(b_0) = \sigma^2 \left[\frac{1}{n} + \frac{\bar{x}_1^2}{\sum (x_{1i} - \bar{x}_1)^2} \right]. \tag{1.5}$$

An unbiased estimate of σ^2 is s^2 given as

$$s^2 = \frac{\sum (y_i - b_0 - b_1 x_{1i})^2}{n - 2}. \tag{1.6}$$

Replacing σ^2 by s^2 in (1.4) and (1.5) we get unbiased estimates of the variances of b_0 and b_1.

Corresponding to the ith observation, the response value predicted by the model is given as

$$\hat{y}_i = b_0 + b_1 x_{1i}. \tag{1.7}$$

The residual corresponding to the ith observation is

$$e_i = y_i - \hat{y}_i \tag{1.8}$$

and the standardized value of the ith residual is defined by

$$e_{is} = \frac{e_i}{s}, \tag{1.9}$$

where s is obtained from (1.6).

In order to construct confidence intervals and to perform tests of hypotheses about the parameters in the regression model, we have to make an additional assumption about the probability law of the u's. The u's are assumed to have a normal distribution.

With this assumption of normality, the least squares estimate, b_1, of β_1 is normally distributed with mean β_1 and variance as given in (1.4). To test the null hypothesis $H_0(\beta_1 = \beta_1^0)$, where β_1^0 is a constant chosen by the investigator, the appropriate test statistic is

$$t = \frac{(b_1 - \beta_1^0)}{\text{s.e.}(b_1)}, \tag{1.10}$$

where s.e.(b_1) is the standard error of b_1 and is given by

$$\text{s.e.}(b_1) = \frac{s}{\left\{ \sum (x_{1i} - \bar{x}_1)^2 \right\}^{1/2}}. \tag{1.11}$$

The statistic t in (1.10) is distributed as a Student's t with $(n-2)$ degrees of freedom. The test is carried out by comparing the observed value with the appropriate tabulated critical t value. The usual test is for $\beta_1^0 = 0$ in which case t reduces to the ratio of b_1 to its standard error.

The confidence limits for β_1 with confidence coefficient $(1 - \alpha)$ are given by

$$b_1 \pm t\left(n-2, \frac{\alpha}{2}\right)[\text{s.e.}(b_1)], \tag{1.12}$$

where $t(n-2, \alpha)$ is the $(1 - \alpha)$ percentile of a t distribution with $(n-2)$ degrees of freedom. The intercept of the regression line, b_0, is normally distributed with mean β_0 and variance given in (1.5). The statistic for testing $H_0(\beta_0 = \beta_0')$, where β_0' is a value specified by the investigator, is

$$t = \frac{b_0 - \beta_0'}{\text{s.e.}(b_0)}, \tag{1.13}$$

where

$$\text{s.e.}(b_0) = s\left[\frac{1}{n} + \frac{\bar{x}_1^2}{\sum (x_{1i} - \bar{x}_1)^2} \right]^{1/2} \tag{1.14}$$

and is distributed as a Student's t with $(n-2)$ degrees of freedom. The confidence limits for β_0 with confidence coefficient $(1 - \alpha)$ are

$$\left\{ b_0 \pm t\left(n-2, \frac{\alpha}{2}\right)[\text{s.e.}(b_0)] \right\}. \tag{1.15}$$

1.4. INDEX OF FIT

After obtaining estimates of β_0, β_1 and σ^2, it is desirable to evaluate the goodness of fit of the model in Equation (1.1) to the observed data. The index most widely used for this purpose is the sample correlation coefficient computed for y and $\hat{y}$, defined* as

$$R = \frac{\sum (y_i - \bar{y})(\hat{y}_i - \bar{\hat{y}})}{\left[\sum (y_i - \bar{y})^2 \sum (\hat{y}_i - \bar{\hat{y}})^2 \right]^{1/2}}, \qquad (1.16)$$

where $\bar{\hat{y}}$ is the average of the $\hat{y}$'s. The numerical value of R lies between 1 and -1. This goodness of fit index may be viewed as a measure of the strength of the linear relationship between y and x_1. The square of the correlation coefficient, R^2 may be written as

$$R^2 = 1 - \frac{\sum (y_i - \hat{y}_i)^2}{\sum (y_i - \bar{y})^2}. \qquad (1.17)$$

The definitions of R given in (1.16) and (1.17) are algebraically equivalent. The definition given in (1.17) provides us with an alternative interpretation. The index R^2 may be interpreted as the proportion of total variability in y that is explained by x_1. If R^2 is near 1, then x_1 explains a large part of variation in y. To examine whether x_1 explains a significant amount of variation in y, the null hypothesis tested is $H_0(\rho = 0)$ against an alternative $H_0(\rho \neq 0)$ where ρ is the population correlation coefficient. The appropriate statistic for testing this hypothesis is

$$t = \frac{|R|\sqrt{n-2}}{\sqrt{1 - R^2}}, \qquad (1.18)$$

where t is a Student's variable with $(n-2)$ degrees of freedom. The test is carried out by comparing the observed t value with a tabulated t value with appropriate degrees of freedom.

It is clear that if no linear relationship exists between y and x_1, then β_1, the population regression coefficient, is zero. Consequently, the statistical tests for $H_0(\beta_1 = 0)$ and $H_0(\rho = 0)$ should be identical. Although the statistics for testing these hypotheses given in (1.10) and (1.18) look different, it can be demonstrated that they are algebraically equivalent.

---

*The numerical value of R may also be obtained by computing the correlation coefficient between y and x_1.

1.5. PREDICTED VALUES AND STANDARD ERRORS

The fitted regression equation can be used to predict the value of the dependent variable y which corresponds to any chosen value, x_1^0, of the independent variable. The predicted value $\hat{y}_0$ is

$$\hat{y}_0 = b_0 + b_1 x_1^0 \qquad (1.19)$$

and has a variance,

$$\text{Var}(\hat{y}_0) = \sigma^2 \left[1 + \frac{1}{n} + \frac{\left(x_1^0 - \bar{x}_1\right)^2}{\sum \left(x_{1i} - \bar{x}_1\right)^2} \right]. \qquad (1.20)$$

An estimate of the variance of $\hat{y}_0$ is obtained by replacing σ^2 by s^2 in (1.20). The confidence limits for the predicted value with confidence coefficient $(1 - \alpha)$ are $\hat{y}_0 \pm t(n - 2, \alpha/2)\text{s.e.}(\hat{y}_0)$, where

$$\text{s.e.}(\hat{y}_0) = s \left[1 + \frac{1}{n} + \frac{\left(x_1^0 - \bar{x}_1\right)^2}{\sum \left(x_{1i} - \bar{x}_1\right)^2} \right]^{1/2}. \qquad (1.21)$$

The predicted response has a normal distribution with mean, $\mu_0 = \beta_0 + \beta_1 x_1^0$. If our interest is in the mean response, then μ_0 is estimated as

$$\hat{\mu}_0 = b_0 + b_1 x_1^0 \qquad (1.22)$$

with variance

$$\text{Var}(\hat{\mu}_0) = \sigma^2 \left[\frac{1}{n} + \frac{\left(x_1^0 - \bar{x}_1\right)^2}{\sum \left(x_{1i} - \bar{x}_1\right)^2} \right]. \qquad (1.23)$$

Note that the point estimate of μ_0 is identical to the predicted response, $\hat{y}_0$. The difference in interpretation is reflected in the variances of the respective quantities.

1.6. EVALUATING THE FIT

We have stated the basic results that are used for making inferences in the context of the simple linear regression model. The results are based on summary statistics that are computed from the data. The results are valid and have meaning only insofar as the assumptions concerning the residual

Table 1.1. Four data sets having same values of summary statistics

	X1	Y1	X2	Y2	X3	Y3	X4	Y4
001	10	8.04	10	9.14	10	7.46	8	6.58
002	8	6.95	8	8.14	8	6.77	8	5.76
003	13	7.58	13	8.74	13	12.74	8	7.71
004	9	8.81	9	8.77	9	7.11	8	8.84
005	11	8.33	11	9.26	11	7.81	8	8.47
006	14	9.96	14	8.10	14	8.84	8	7.04
007	6	7.24	6	6.13	6	6.08	8	5.25
008	4	4.26	4	3.10	4	5.39	19	12.50
009	12	10.84	12	9.13	12	8.15	8	5.56
010	7	4.82	7	7.26	7	6.42	8	7.91
011	5	5.68	5	4.74	5	5.73	8	6.89

Source: Anscombe (1973).

terms in the model are satisfied. Consequently, it is very important to investigate the structure of the residuals and the data pattern through graphs. A large value of R^2 or a significant t statistic does not insure that the data has been fitted well. To emphasize this point, Anscombe (1973) has constructed four data sets, each with a distinct pattern, but each having the same set of summary statistics. The data and graphs are reproduced in Table 1.1 and Figure 1.1. An analysis based exclusively on an examination of summary statistics would have been unable to detect the differences in patterns, thereby producing an incorrect analysis.

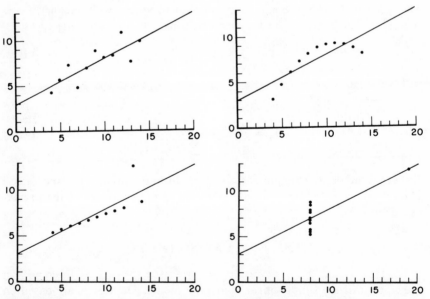

Fig. 1.1.   Plot of the data (x, y) with the fitted line for four data sets (Table 1.1). Source: Anscombe (1973).

The method of least squares analysis is quite robust in that small or minor violations of the underlying assumptions do not invalidate the inferences or conclusions drawn from the analysis in a major way. Gross violations of the model assumptions can, however, seriously distort conclusions. In the next section we outline an approach for assessing the validity of the assumptions. We reemphasize that the prime focus of this book is on the detection and correction of violations of the basic linear model assumptions as a means of achieving a thorough and informative analysis of the data.

1.7. ANALYSIS OF RESIDUALS

A simple and effective method for detecting model deficiencies in regression analysis is by examining the residuals. The ith residual has been defined as

$$e_i = y_i - \hat{y}_i.$$

Corresponding to e_i, we have also defined the ith standardized residual e_{is},

$$e_{is} = \frac{e_i}{s},$$

where s is the standard deviation of residuals as given in Equation (1.6).

The standardized residuals e_{is} have zero mean and unit standard deviation. With a moderately large sample, these residuals should be distributed approximately as independent, normal deviates. The residuals are not strictly independently distributed, but with a large number of observations, the lack of independence may be ignored. An appropriate graph of the residuals will often expose gross model violations when they are present. Studying residual plots is one of the main tools in our analysis. We examine several different types of plots looking for various different model violations. Rather than producing a catalog here of all the different residual plots that may be employed, we describe them in the examples where they are used. Some of the more commonly used plots are those in which the standardized residuals e_{is} are plotted as the ordinate against

1. the fitted value, $\hat{y}$
2. the independent variable, x_1
3. the time order in which the observations occur

as abscissa* (Draper and Smith, 1966; Anscombe, 1973). In general, when

---

*The assumption that the residuals are approximately normally distributed may be verified by constructing a normal probability plot (Daniel and Wood, 1971). We have not included any discussion of this plot. Regression theory is robust with respect to misspecification of the probability law of the residuals. Other model violations that may be detected using a normal probability plot are more directly identified in the plots that we have just listed.

the model is correct, the standardized residuals tend to fall between 2 and
-2 and are randomly distributed about zero. The residual plots should
show no distinct pattern of variation. When the model is invalid, residual
plots have characteristics that are different from those just described. We
emphasize that an essential part of any regression analysis includes a
careful examination of residuals to ensure that the assumptions of least
squares theory are in order. The process of checking for gross model
violations by analyzing residuals is a very useful exercise for uncovering
hidden structures in the data. These ideas are explicitly demonstrated in
the various examples throughout the book.

1.8.   REPAIR TIMES FOR COMPUTERS

This first example is used to illustrate some of the preceding standard
statistical results. We are concerned with a company that markets and
repairs small computers. The company wants to forecast the number of
service engineers that will be required over the next few years. One element
of the forecasting procedure is an analysis of the length of service calls.
The length of a call depends on the number of electronic components in
the computer that must be repaired or replaced. To establish this relationship a sample of records on service call was taken. The data consists of the
length of service calls in minutes and the number of components repaired.
The data is presented in Table 1.2. The graph of response versus explanatory variable in Figure 1.2 suggests a straight line relationship.

A linear model

$$\text{MINUTES} = \beta_0 + \beta_1 \cdot \text{UNITS} + u$$

is fitted to the data. Table 1.3 gives the estimated coefficients and their

Table 1.2.   Length of service calls and the number of units repaired

UNITS	MINUTES
1	23
2	29
3	49
4	64
4	74
5	87
6	96
6	97
7	109
8	119
9	149
9	145
10	154
10	166

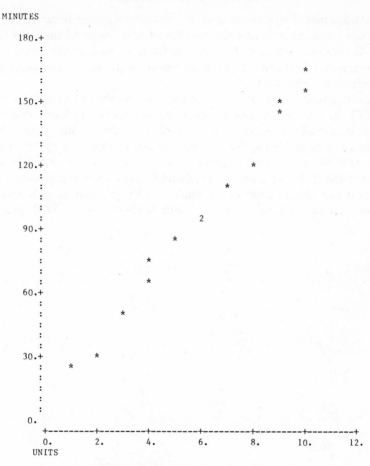

MINUTES

```
180.+
    :
    :
    :                                                    *
    :
150.+                                            *
    :                                            *
    :
    :
    :
120.+                                    *
    :                              *
    :
    :
    :                          2
 90.+
    :                    *
    :              *
    :
    :              *
 60.+
    :        *
    :
    :
    :
 30.+      *
    :   *
    :
    :
    :
    :
  0.
    +---------+---------+---------+---------+---------+---------+
    0.        2.        4.        6.        8.       10.       12.
UNITS
```

Fig. 1.2. Plots of minutes versus units.

**Table 1.3.   Estimates of regression coefficients and
standard errors**

Variable	Coefficient	SE[a]	t
UNITS	15.509	.505	30.71
CONSTANT	4.162	3.355	1.24
$n = 14$	$R^2 = .987$	$s = 5.392$	

[a] Standard error.

11

standard errors. The high value of R^2 indicates a strong linear relationship between servicing time and the number of units repaired during a service call. Before proceeding with further analysis we look at the residual plots to ensure that there are no serious violations of the underlying assumptions associated with the model.

The standardized residuals are plotted against the independent variable, UNITS, in Figure 1.3. The residuals appear to be randomly distributed about 0 and all lie between ± 2. There is no discernible pattern to the distribution of residuals; that is, they do not change in a systematic way with UNITS. A systematic pattern of variation of the residuals would indicate either one or more inadequacies in the underlying assumptions or errors in the specification of the equation. These deficiencies would have to be corrected before proceeding with further analysis. With plots like

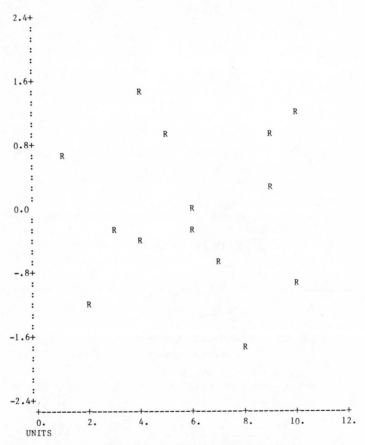

Fig. 1.3.   Plot of standardized residuals against units.

those in Figure 1.3 we conclude that the model specification is satisfactory, and proceed with the analysis.

We begin by examining the adequacy of the explanatory variable. The correlation coefficient has already indicated that nearly 99% of the variation in the length of service calls can be explained by the number of components repaired. We can formally assess the explanatory ability of UNITS by testing $H_0(\beta_1 = 0)$, where β_1 denotes the regression coefficient of MINUTES on UNITS. The appropriate statistic from (1.10) is

$$t = \frac{15.509}{.505} = 30.71, \quad \text{d.f.} = 12.$$

The observed value of t is highly significant, leading to the rejection of the null hypothesis.

The fitted equation is

$$\text{MINUTES} = 4.162 + 15.509 \cdot \text{UNITS}.$$

The coefficients in the above equation can be interpreted in simple physical terms. The constant term represents the set up or start up time for each repair and is approximately 4 min. The coefficient of UNITS represents the increase in the length of a service call for each additional component that has to be repaired. From the given data, we estimate that it takes about 16 min (15.509) for each additional component that has to be repaired. For illustration let us suppose that management expected the increase in service time for each additional unit to be repaired to be 12 min. Does the data support this conjecture? The answer may be obtained by testing $H_0(\beta_1 = 12)$ against $H_1(\beta_1 \neq 12)$. The appropriate statistic is

$$t = \frac{b_1 - 12}{\text{s.e.}(b_1)} = \frac{15.509 - 12}{.505} = 6.948, \quad \text{d.f.} = 12.$$

The result is highly significant leading to the rejection of the null hypothesis. Management's estimate of the increase in time for each additional component to be repaired is not supported by the data. Their estimate is too low.

We shall next construct a confidence interval for β_1. From Equation (1.12) it follows that the 95% confidence limits for β_1 are ($15.509 \pm 2.18 \times 0.505$), which simplifies to (14.408, 16.610). That is, the incremental time required for each broken unit is between 14 and 17 min.

Suppose we wanted to predict the length of service call in which four components had to be repaired. If $\hat{y}_4$ denotes the predicted value, then from (1.19) we get

$$\hat{y}_4 = 4.162 + 15.509(4) = 66.198$$

with a standard error which is obtained from (1.21) and is

$$\text{s.e.}(\hat{y}_4) = 5.392 \left[1 + \frac{1}{14} + \frac{(4-6)^2}{114} \right]^{1/2} = 5.672.$$

On the other hand if the service department wanted to estimate the expected service time for a call which needed four components to be replaced, we would use (1.22) and (1.23), respectively. Denoting the expected service time for a call which needed four components to be fixed by μ_4, we have from (1.22) and (1.23)

$$\hat{\mu}_4 = 4.162 + 15.509 \times 4 = 66.198,$$

$$\text{s.e.}(\hat{\mu}_4) = 5.392 \left(\frac{1}{14} + \frac{(4-6)^2}{114} \right)^{1/2} = 1.759.$$

With these standard errors we can construct appropriate confidence intervals.

As can be seen from Equation (1.21) the standard error of prediction increases the further the value of the independent variable is from the center of the actual observations. Care should be taken when predicting the value of MINUTES corresponding to a value for UNITS which does not lie close to the observed data. There are two dangers in such predictions. There is substantial uncertainty due to the large standard error. More importantly, the linear relationship that has been estimated may not hold outside the range of observations. Care should therefore be taken in employing fitted regression lines for prediction far outside the range of observations. In our example, we should not use the fitted equation to predict the service time for a service call which requires 25 components to be replaced or repaired. This value lies too far outside the existing range of observations.

In summary, we believe that the 14 data points have given us an informative view of the repair time problem. There is no evidence that the underlying assumptions of regression analysis are not in order. Within the range of observed data, we are confident of the validity of our inferences and predictions.

As a preview to the ensuing chapters, and as a means of placing further emphasis on the necessity for looking at both summary statistics and data patterns, we present some additional data on the computer repair problem discussed above. In a second sampling period, 10 more observations on the variables MINUTES and UNITS were obtained. These observations happened to be concentrated at larger values of UNITS than the original sample. Since all observations were collected by the same method from a fixed environment, all 24 observations were pooled to form one data set. The data appears in Table 1.4 and is plotted in Figure 1.4. The graph suggests that a linear model may still be a good starting point, but there appears to be some curvature beginning around UNITS = 13 or 14. The summary statistics found in Table 1.5 have values that are different from those obtained with the initial sample. R^2 is still sizable (.900) and the t value is significant, implying a good fit to the data. However, the plot of residuals in Figure 1.5 clearly shows that there is a problem. The residuals are not randomly distributed about zero. In fact, they systematically move from negative to positive and back to negative as the value of units increases. The graph suggests that there is some information left in the UNITS variable that has not been extracted in the linear equation. The problem has to do with the levelling of the MINUTES variable as UNITS increased beyond 14. This effect has not been captured by the proposed linear equation, but the residuals have clearly indicated the problem. The analysis up to this point suggests that there is an added element of time efficiency in repairing a computer when the number of units in need of repair or replacement exceeds 13 or 14. The analyst in this study can now

Table 1.4.   Length of service calls and the number of units repaired in expanded sample

UNITS	MINUTES
1	23
2	29
3	49
4	64
4	74
5	87
6	96
6	97
7	109
8	119
9	149
9	145
10	154
10	166
11	162
11	174
12	180
12	176
14	179
16	193
17	193
18	195
18	198
20	205

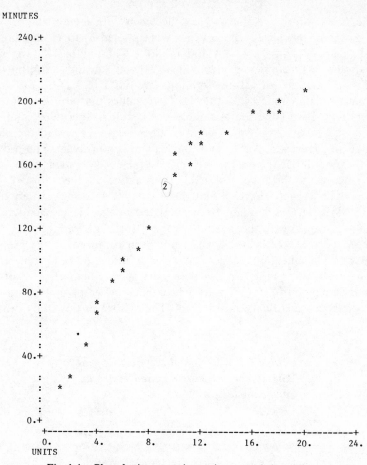

Fig. 1.4. Plot of minutes against units, expanded sample.

**Table 1.5.   Regression coefficients and standard
errors (expanded sample)**

Variable	Coefficient	SE	t
UNITS	9.969	.722	13.81
CONSTANT	37.213	7.985	4.66
$n = 24$	$R^2 = .900$	$s = 18.753$	

16

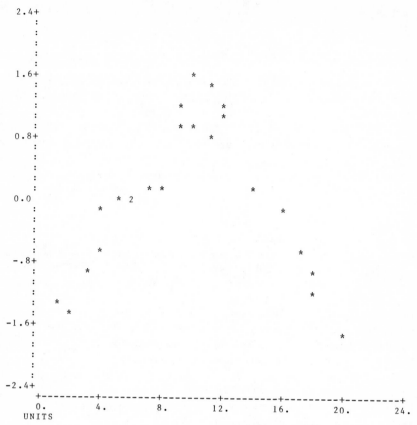

Fig. 1.5.   Residuals versus units, expanded sample.

go back to the service division of the company to further investigate the reasons behind this apparent efficiency.

The task of finding a better model to fit the data is left to the reader. We suggest that a new analysis may be attempted after the reading of Chapters 2 and 3.

We end this chapter with two final points. First, the deficiency in the linear equation is initially suggested by the graph of the raw data in Figure 1.4. However, the inadequacy of the model is magnified in the plot of residuals. As a general rule the residual plots are more informative about model deficiencies than graphs of the raw data. Second, extrapolation beyond the range of the data can be misleading as is evident from a comparison of the results obtained from the original and expanded data sets.

BIBLIOGRAPHIC NOTES

The standard theory of regression analysis is developed in a number of good texts. Some of these texts have been written to serve specific disciplines. Each of them provides a complete treatment of the standard results. The books by Brownlee (1965), Snedecor and Cochran (1967), and Kmenta (1971) develop the results using simple algebra and summation notation. The development in Johnston (1972), Rao (1965), and Searle (1971) is more terse and leans heavily on matrix algebra.

REFERENCES

Anscombe, F. J., Graphs in statistical analysis, *American Statistician*, **27**, 17–21 (1973).

Beaton, A. E., D. B. Rubin, and J. L. Barone, The acceptability of regression solutions: another look at computational accuracy, *J. Amer. Stat. Assoc.*, **71**, 158–168 (1976).

Brownlee, K. A., *Statistical Theory and Methodology in Science and Engineering*, Wiley, New York, 1965.

Daniel, C. and F. S. Wood, *Fitting Equations to Data*, Wiley, New York, 1971.

Draper, N. R. and H. Smith, *Applied Regression Analysis*, Wiley, New York, 1966.

Johnston, J., *Econometric Methods*, McGraw-Hill, New York, 1972.

Kmenta, J., *Elements of Econometrics*, Macmillan, New York, 1971.

Longley, J. W., An appraisal of least square programs for the electronic computers from the point of view of the user, *J. Amer. Stat. Assoc.*, **62**, 819–841 (1967).

Rao, C. R., *Linear Statistical Inference and its Applications*, Wiley, New York, 1973.

Searle, S. R., *Linear Models*, Wiley, New York, 1971.

Snedecor, G. W. and W. G. Cochran, *Statistical Methods*, Iowa State University Press, Ames, Iowa, 1967.

CHAPTER 2

Detection and Correction of Model Violations: Simple Linear Regression

2.1. INTRODUCTION

In this chapter we consider two problems that may arise in regression analysis with one independent variable. The same problems may also arise with more than one independent variable but we defer that discussion to later chapters.

The first problem discussed is the presence of extreme data points or outliers in a set of observations. When analyzing a body of data it is necessary to make sure that conclusions are not solely dependent on one or two extreme observations. Outliers may occur because of gross errors during the recording or transcribing of the data. On the other hand, they may also be genuine observations, highly significant and suggestive, to which we should pay considerable attention. Whichever be the case, extreme data points should always be followed up and examined in detail. We outline a method for detecting extreme observations and studying their implications.

The second problem considered is that of transformation of data. For a variety of reasons, it often becomes necessary to fit a linear regression model to transformed rather than original variables. We discuss the situations where it is necessary to transform the data, the possible choices of transformation, and the analysis of transformed data.

2.2. EFFECTS OF OUTLIERS IN SIMPLE REGRESSION

In this section we investigate the notion of outliers or extreme data points and discuss how these observations may affect the regression results. Outliers are data points with large residuals. Observations classified as outliers have residuals that are large relative to the residuals for the remainder of the observations.

We assume that the relationship between two variables, x and y, is well approximated by a straight line. The formal regression model is

$$y_i = \beta_0 + \beta_1 x_i + u_i, \tag{2.1}$$

where β_0 and β_1 are parameters to be estimated. The u's are unobservable random errors that are assumed to be normally, independently distributed with zero mean and constant variance σ^2. Given sufficient data on x and y the model parameters can be estimated by the least squares method. Interpretation of the regression coefficients is immediate provided that there are no serious violations of the basic underlying regression assumptions. The assumptions are checked by analyzing the regression residuals.

The analysis of residuals usually begins with graphs of the residuals against the independent variable x, and the fitted value $\hat{y}$. In general, these graphs are starting points for checking the deficiencies of model assumptions, such as inadequacy of straight line assumption, lack of constant variance, presence of outliers, and correlated errors. We presently restrict our attention to detecting outliers and gauging their effects on the regression results. The following illustrative example is taken from the field of commercial broadcasting.

2.3. TELEVISION RATING DATA

The success of a program on commercial television is determined in part by a rating system that is an attempt to measure the show's ability to attract and hold viewers. In real terms, the rating points generate sponsor interest and in turn bring revenue to the station. A station manager concerned about news show ratings contracted for a study to identify the factors that affect the ratings. In addition to the obvious variables such as format, special features, and individual newscaster appeal, it was suggested that there was a holdover effect from the show immediately preceding the news. The rating of the news show was partially dependent on the rating of the show that preceded it, that is, the "lead-in" show. In order to quantify this effect, a random sample of previous ratings was taken across regions and for various points in time over the past 2 years. It is assumed that factors other than lead show rating would average out over the sample. The data consists of observations on a variable denoted as y, news show rating, and a second variable called x representing lead show rating. The data is displayed in Table 2.1. The rating is an index that ranges between 1

Table 2.1. Lead rating (x) and news rating (y) of television data

ROW	X	Y
* 1 *	2.50	3.80
* 2 *	2.70	4.10
* 3 *	2.90	5.80
* 4 *	3.10	4.80
* 5 *	3.30	5.70
* 6 *	3.50	4.40
* 7 *	3.70	4.80
* 8 *	3.90	3.60
* 9 *	4.10	5.50
* 10 *	4.30	4.15
* 11 *	4.50	5.80
* 12 *	4.70	3.80
* 13 *	4.90	4.75
* 14 *	5.10	3.90
* 15 *	5.30	6.20
* 16 *	5.50	4.35
* 17 *	5.70	4.15
* 18 *	5.90	4.85
* 19 *	6.10	6.20
* 20 *	6.30	3.80
* 21 *	6.50	7.00
* 22 *	6.70	5.40
* 23 *	6.90	6.10
* 24 *	7.10	6.50
* 25 *	7.30	6.10
* 26 *	7.50	4.75
* 27 *	2.50	1.00
* 28 *	2.70	1.20
* 29 *	7.30	9.50
* 30 *	7.50	9.00

and 10. Approximating the relationship between y and x as a straight line we have

$$y_i = \beta_0 + \beta_1 x_i + u_i. \tag{2.2}$$

Analysis

The first step is to plot the data, y versus x as in Figure 2.1. The graph indicates an upward sloping straight line relationship. There is also an indication that outliers are present. We proceed by estimating the parameters of the model. At first glance the regression results (Table 2.2) are impressive. A report to the station manager would emphasize that almost 40% of the variation in news show ratings can be explained by the rating of the lead show. Furthermore, for each rating point increase in the lead show, the news show rating is expected to increase by .66. Or using a confidence interval we could report that with 95% confidence the interval

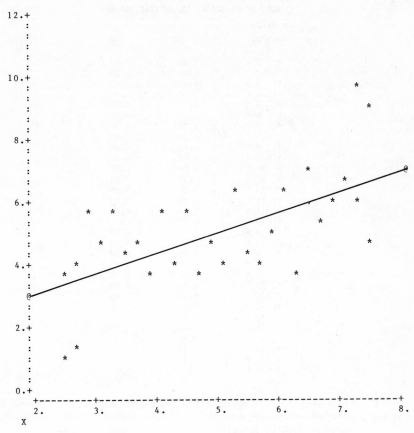

Fig. 2.1. Graph of news rating (y) against lead rating (x).

Table 2.2. Regression coefficients and standard errors for television rating data

Variable	Coefficient	SE	t
x	.665	.155	4.287
CONSTANT	1.706	.817	2.088
$n = 30$	$R^2 = .396$	$s = 1.402$	

22

that ranges approximately from .355 to .975 covers the true marginal contribution of an additional lead show rating point. These statements are correct provided that the model is correct, that is provided there are no serious violations of the assumptions underlying the analysis. As we shall see below, the data does contain outliers and they have a notable effect on the regression results.

2.4. MODEL ADEQUACY AND RESIDUAL PLOTS

Our basic concern now is to examine the adequacy of the straight line approximation. The investigation proceeds by constructing several graphs. We inspect the observed standardized residuals against the independent

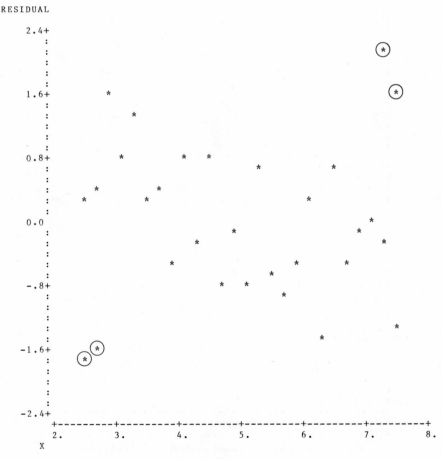

Fig. 2.2. Graph of standardized residuals against (x).

variable (x) (Figure 2.2), and the standardized residuals against fitted values ($\hat{y}$) (Figure 2.3). From the examination of the graphs, it is clear that the straight line assumption is not confirmed. From Figures 2.2 and 2.3 we see that there are four points, two at each extreme (marked by circles) which have large residuals. For this data both plots (Figures 2.2 and 2.3) give the same information. From Figure 2.2 we see that within the middle range of x the graph looks acceptable, that is, the residuals appear to be randomly distributed about $e = 0$. However, for smaller values of x most values of e are positive which indicates predictions by the model are lower than the observed value. Furthermore, there is a counterbalance of two large negative e's suggesting excessive overprediction by the model. For

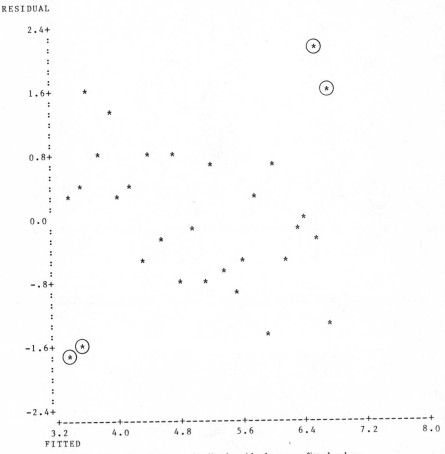

Fig. 2.3. Graph of standardized residuals versus fitted values.

large values of x the picture is exactly reversed. We conclude that there appear to be four points that do not belong with the rest of the data. We can anticipate the effect that these four points alone have on the regression results by looking at all data points plotted about the estimated regression line (Figure 2.1). It is fairly clear that the four points inflated the slope of the line. These four points should be regarded as outliers, and thoroughly investigated. In fact it appears that the slope determined by the data excluding the four points may be very close to zero. In other words, y is not affected by x; that is, news show ratings (y) do not seem to be influenced by the rating of the "lead-in" show (x).

2.5. DELETION OF DATA POINTS

The conjecture can be checked by dropping the four points from the data and reestimating the regression equation. As seen from the results in Table 2.3 and the residual plot in Figure 2.4, the claim is substantiated. The marginal effect of x is diminished and the proportion of explained variation is decreased. The residuals appear to be randomly distributed about the line $e = 0$ indicating that the proposed equation (2.2) is a satisfactory model for analyzing the rating data, after the deletion of the four points.

But what of the four data points that were deleted? It is possible that their extreme behavior is a result of measurement or transcription errors in which case they should be deleted and forgotten. However, if that is not the case, the outliers may provide more valuable information about the program rating system and relationships between y and x than the well-behaved observations that are neatly explained by the regression equation. Table 2.4 gives the summary statistics for both the full and reduced data sets. In fact, the result derived from the well-behaved observations is that x has little effect on y. Therefore, it may be most valuable to try to understand the special circumstances that generated the extreme responses.

The estimated values of regression coefficients and supporting summary statistics are very sensitive to outliers. When there are extreme data points present, such as in the television rating data, it is possible that a very small

Table 2.3. Regression coefficients and standard errors for reduced data set

Variable	Coefficient	SE	t
x	.260	.121	2.147
CONSTANT	3.713	.631	5.887
$n = 26$	$R^2 = .161$	$s = 0.925$	

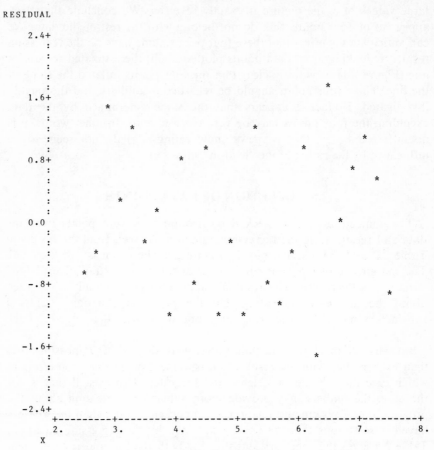

Fig. 2.4. Graph of standardized residuals versus $x(n=26)$.

Table 2.4. Summary of regression results for the full and reduced data sets

	Full data set	Reduced data set
b_1	0.665	0.260
b_0	1.706	3.713
R^2	0.396	0.161
s	1.402	0.925
n	30	26

proportion of the data has a very large effect on the regression equation. A researcher may or may not find that imbalance acceptable. What is of more importance is that he or she must know how the data appears. Graphs of the residuals provide the necessary information. The importance of any subset of data can then be evaluated empirically by computing the regression equation on data that first includes and then excludes the questionable subset. As Anscombe (1973) says, "We are usually happier about asserting a regression relation if the relation is still appropriate after a few observations (any ones) have been deleted—that is, we are happier if the regression relation seems to permeate all the observations and does not derive largely from one or two."

Formal statistical decision rules for detecting outliers have been proposed by Anscombe (1960), Anscombe and Tukey (1963), Rosner (1975), Ellenberg (1976). These rules are rigid in comparison to the empirical approach described previously. The rules are usually defined to reject an observation if the standardized residual corresponding to the observation exceeds a certain multiple of the mean square error. The formal rules should be applied with some caution. Most of these procedures would not cast suspicion on any of the observations in the news rating analysis. There, the largest residuals are approximately two standard deviations from the mean of zero. It is a combination of the actual magnitude and the pattern of residuals that suggests problems. As with the television rating problem, an analysis based on residual plots and possible deletion of some observations may lead to conclusions that are quite different from those conclusions based only on summary statistics. In addition, those observations that give rise to large residuals should be investigated. They may bring to light measurement or transcription errors and serve the function of data editing, but they also may point to some significant feature of the data which may be glossed over or missed in a broad overall analysis.

2.6. TRANSFORMATION OF VARIABLES

A convenient starting point in regression analysis is that the model describing the data is linear in the variables. In order to accomplish this, the analysis is frequently carried out on transformed variables. The necessity for transforming the data arises because the original variable, or the model in terms of the original variable, violates one or more of the standard assumptions. The most commonly violated assumptions are those concerning the linearity of the model and the constancy of the error variance. A regression model is linear (as has already been remarked in Chapter 1) when the parameters present in the model occur linearly. Each of the four following models is linear:

$$y = \beta_0 + \beta_1 x + u$$

$$y = \beta_0 + \beta_1 x + \beta_2 x^2 + u$$

$$y = \beta_0 + \beta_1 \log x + u$$

$$y = \beta_0 + \beta_1 \sqrt{x} + u$$

because the model parameter β's enter linearly. On the other hand

$$y = \beta_0 + e^{\beta_1 x} + u$$

is a nonlinear model because the parameter β_1 does not enter the model linearly. To satisfy the assumptions of the standard regression model, instead of working with the original variables, we sometimes work with transformed variables. Transformations may be necessary for several reasons. These may be summarized as follows:

   a.   Theoretical considerations may specify that the relationship between two variables is nonlinear. An appropriate transformation of the variables can make the relationship between the transformed variables linear. Consider an example from learning theory (experimental psychology). A learning model that is widely used states that the time taken to perform a task on the ith occasion (T_i) is

$$T_i = \alpha \beta^i, \qquad \alpha > 0, \qquad 0 < \beta < 1. \tag{2.3}$$

       The relationship between T_i and i as given in (2.3) is nonlinear, and we cannot directly apply techniques of linear regression. On the other hand, if we take logarithms of both sides we get

$$\log T_i = \log \alpha + i \log \beta \tag{2.4}$$

       showing that $\log T_i$ and i are linearly related. The transformation enables us to use standard regression methods. Although the relationship between the original variables was nonlinear, the relationship between transformed variables is linear. A transformation is used to achieve the linearity of the fitted model.

   b.   The dependent variable y, which is analyzed, may have a probability distribution whose variance is related to the mean. If the mean is related to the value of the independent variable x, then the variance of y will change with x, and the variance will not be constant. The distribution of y will usually also be nonnormal under these conditions. Nonnormality invalidates the standard tests of significance (although not in a major way with large samples) since they are based on the normality assumption. The unequal variance of the error terms will produce estimates that are unbiased, but are no longer best in the sense of precision (variance). In these situations we often transform the data so as to ensure normality and constancy of error variance. In practice, the transformations are chosen to ensure the constancy of variance (variance stabilizing transformations). It is a fortunate coincidence that the variance stabilizing transformations are also good normalizing transforms.

c.   There are neither prior theoretical nor probabilistic reasons to suspect that a transformation is required. The evidence comes from examining the residuals from the fit of a simple linear model.

Each of these cases is illustrated in the following sections.

2.7.   TRANSFORMATIONS TO ACHIEVE LINEARITY

One of the standard assumptions made in regression analysis is that the model which describes the data is linear. From theoretical considerations, or from an examination of scatter plot of y against x, the relationship between y and x may appear to be nonlinear. There are, however, several nonlinear models which by appropriate transformations can be made linear. We list some of these linearizable curves in Table 2.5. The corresponding graphs are given in Figure 2.5.
When curvature is observed in the plot of y against x, a linearizable curve from one of those given in Figure 2.5 may be chosen to represent the data.

Table 2.5.   Linearizable functions with corresponding transformations

Function	Transformation	Linear form	Graph shown in Figure
$y = \alpha x^\beta$	$y' = \log y, \quad x' = \log x$	$y' = \log \alpha + \beta x'$	2.5a, b
$y = \alpha e^{\beta x}$	$y' = \ln y$	$y' = \ln \alpha + \beta x$	2.5c, d
$y = \alpha + \beta \log x$	$x' = \log x$	$y = \alpha + \beta x'$	2.5e, f
$y = \dfrac{x}{\alpha x - \beta}$	$y' = \dfrac{1}{y}, \quad x' = \dfrac{1}{x}$	$y' = \alpha - \beta x'$	2.5g, h
$y = \dfrac{e^{\alpha + \beta x}}{1 + e^{\alpha + \beta x}}$ [a]	$y' = \ln \left(\dfrac{y}{1 - y} \right)$	$y' = \alpha + \beta x$	2.5i

[a] In Chapter 5 we describe an application using this transformation.

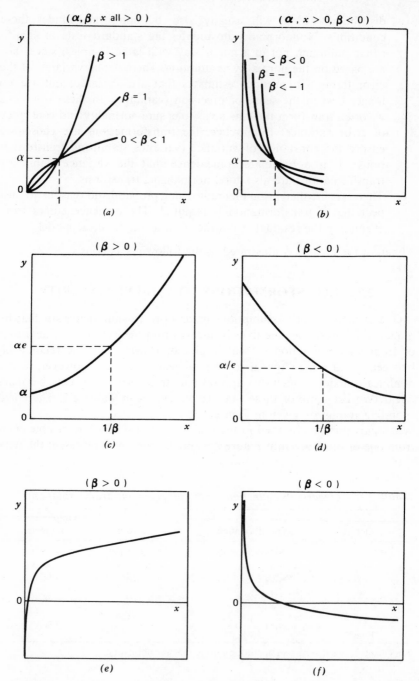

Fig. 2.5 Graphs of Linearizeable Functions. Source: Daniel and Wood, (1971), pp. 20–21.

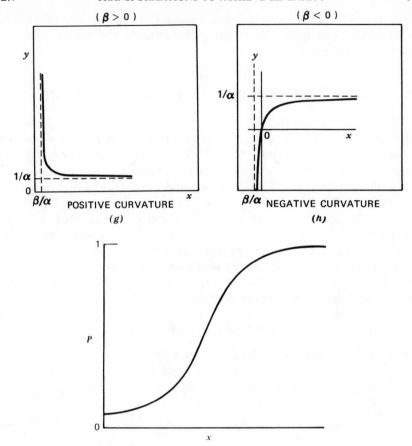

Fig. 2.5.   Graphs of linearizable functions. Source: Daniel and Wood (1971), pp. 20–21.

There are, however, many simple nonlinear models that cannot be linearized. Consider for example,

$$y = \alpha + \beta\delta^x$$

a modified exponential curve, or

$$y = \alpha_1 e^{\theta_1 x} + \alpha_2 e^{\theta_2 x}$$

which is the sum of two exponential functions. The strictly nonlinear models (i.e., those not linearizable by variable transformation) require very different methods for fitting. We shall not describe them in the present text but refer the interested reader to Gallant (1975).

In the following example, theoretical considerations lead to a model that is nonlinear. The model is, however, linearizable and we indicate the appropriate analysis.

2.8.   BACTERIA DEATHS DUE TO X-RAY RADIATION

The data given in Table 2.6 represent the number of surviving bacteria (in hundreds) as estimated by plate counts in an experiment with marine bacterium following exposure to 200 kilovolt X-rays for periods ranging from $t = 1$ to 15 intervals of 6 min. The experiment was carried out to test the single hit hypothesis of X-ray action under constant field of radiation. According to this theory, there is a single vital center in each bacterium, and this must be hit by a ray before the bacteria is inactivated or killed. The particular bacterium studied does not form clumps or chains, so the number of bacterium can be estimated directly from plate counts. If the theory is applicable, the logarithm of the number of survivors should plot as a straight line against length of exposure time (t). If n_t represents the number surviving after exposure time t then

$$n_t = n_0 e^{\beta t}, \qquad t \geqslant 0, \tag{2.5}$$

where n_0 and β are parameters. The parameters n_0 and β have simple physical interpretations; n_0 is the number of bacteria at the start of the experiment, while β is the destruction (decay) rate. Taking logarithms of both sides of (2.5) we get

$$\ln n_t = \ln n_0 + \beta t = \alpha + \beta t, \tag{2.6}$$

Table 2.6.   *Number of surviving bacteria (units of 100)*

	N	TIME
001	355	1
002	211	2
003	197	3
004	166	4
005	142	5
006	106	6
007	104	7
008	60	8
009	56	9
010	38	10
011	36	11
012	32	12
013	21	13
014	19	14
015	15	15

where $\alpha = \ln n_0$, and we have $\ln n_t$ as a linear function of t. If we introduce u_t as the random error, our model becomes

$$\ln n_t = \alpha + \beta t + u_t \tag{2.7}$$

and we can now apply standard least square methods.

In order to get the error u_t in the transformed model (2.7) to be additive, the error must occur in the multiplicative form in the original model (2.5). The correct representation of the model should be

$$n_t = n_0 e^{\beta t} u_t', \tag{2.8}$$

where u_t' is the multiplicative random error. By comparing (2.7) and (2.8) it is seen that $u_t = \ln u_t'$. For standard least square analysis u_t should be normally distributed which in turn implies that u_t' has a log-normal distribution.* In practice after fitting the transformed model we look at the residuals from the fitted model to see if the model assumptions hold. No attempt is usually made to investigate the random component u_t' of the original model.

Inadequacy of a Linear Model

The first step in the analysis is to plot the raw data, n_t versus t. The plot, shown in Figure 2.6 suggests a nonlinear relationship between n_t and t. However, we proceed by fitting the simple linear model and investigate the consequences of misspecification. The model is

$$n_t = \alpha + \beta t + u_t, \tag{2.9}$$

where α and β are constants, u_t's are the random errors with zero expectations, equal variances and uncorrelated with each other. The estimates of α, β, their standard errors, and the square of the correlation coefficient are given in Table 2.7.

Table 2.7. Estimated regression coefficients (n versus time)

Variable	Coefficient	SE	t
TIME (t)	-19.46	2.50	-7.788
CONSTANT	259.58	22.730	11.420
$n = 15$	$s = 41.832$	$R^2 = 0.823$	

*The random variable y is said to have a lognormal distribution if $\ln y$ is normally distributed. See Aitchison and Brown (1957).

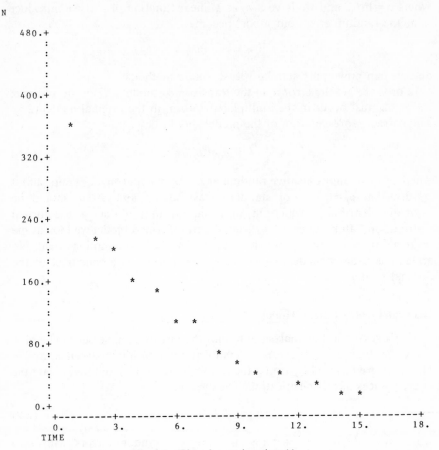

Fig. 2.6.   Plot of n_t against time (t).

In spite of the fact that the regression coefficient for the time variable is significant and we have a high value of R^2, the linear model is not appropriate. The plot of n_t against t shows departure from linearity for high values of t (Figure 2.6). We see this even more clearly if we look at a plot of the residuals against time. In Figure 2.7 the standardized residuals are plotted against time.

The distribution of residuals has a distinct pattern. The residuals for $t = 2$ through 11 are all negative, for $t = 12$ through 15 are all positive, whereas the residual for $t = 1$ appears to be an outlier. This systematic pattern of deviation confirms that the linear model does not fit the data.

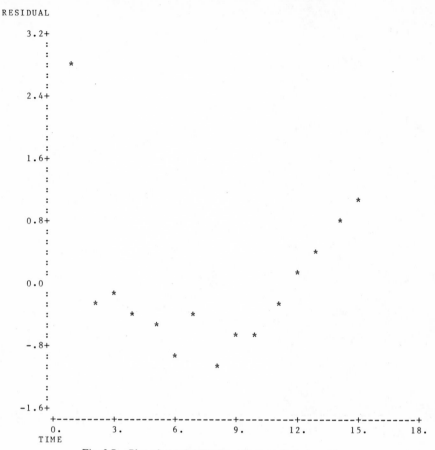

Fig. 2.7.   Plot of standardized residuals against time (t).

Logarithmic Transformation for Achieving Linearity

The relation between n_t and t appears distinctly nonlinear and we will work with the transformed variable $\ln n_t$ which is suggested from theoretical considerations as well as by Figure 2.8. The plot of $\ln n_t$ against t appears linear indicating that the logarithmic transformation is appropriate. The results of fitting Equation (2.7) appear in Table 2.8.

The coefficients are highly significant, the standard errors are reasonable, and nearly 98% of the variation in the data is explained by the model. The standardized residuals are plotted against t in Figure 2.9. There are no systematic patterns to the distribution of the residuals and the plot is

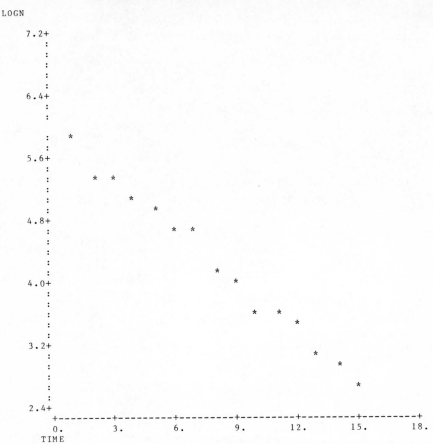

Fig. 2.8. Plot of $\ln n_t$ against time (t).

Table 2.8. Estimated regression coefficients ($\ln n$ versus time)

Variable	Coefficient	SE	t
TIME (t)	−0.218	.00657	−33.22
CONSTANT	5.973	.0598	99.92
$n = 15$	$s = 0.110$	$R^2 = 0.9884$	

RESIDUAL

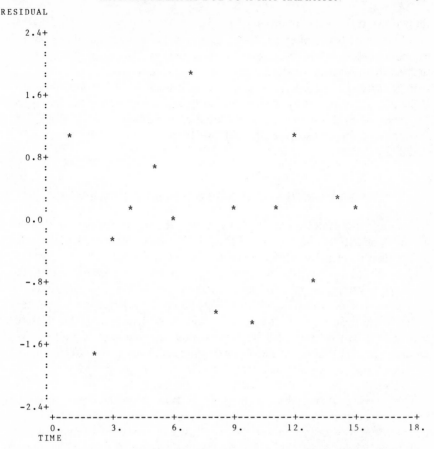

Fig. 2.9. Plot of standardized residuals against time (t) after transformation.

satisfactory. The single-hit hypothesis of X-ray action which postulates that $\ln n_t$ should be linearly related to t is confirmed by the data.

While working with transformed variables, careful attention must be paid to the estimates of the parameters of the model. In our example the point estimate of β is -0.218 and the 95% confidence interval for the same parameter is $(-0.232, -0.204)$. The estimate of the constant term in the equation is the best linear unbiased estimate of $\ln n_0$. If a denotes the estimate, e^a may be used as an estimate of n_0. With $a = 5.973$, the estimate of n_0 is $e^a = 392.68$. This estimate is not an unbiased estimate of n_0; that is, the true size of the bacteria population at the start of the experiment was probably somewhat smaller than 392.68. A correction can be made to reduce the bias in the estimate of n_0. The estimate $\exp(a - \left(\frac{1}{2}\right)$ s.e. $(a))$ is nearly unbiased (Goldberger, 1968). In our present example, the modified estimate of n_0 is 381.11. Note that the bias in estimating n_0 has no effect on the test of the theory or the estimation of the decay rate.

In general, if nonlinearity is present it will show up in a plot of the data. If the plot corresponds approximately to one of the graphs given in Figure 2.5, then one of those curves can be fitted after transforming the data. The adequacy of the transformed model can then be investigated along ways outlined previously.

2.9. TRANSFORMATIONS TO STABILIZE VARIANCE

We have discussed in the previous section the use of transformations to achieve linearity of the regression function. Transformations are also used to stabilize the error variance, that is, to make the error variance constant for all the observations. The constancy of error variance is one of the standard assumptions of least square theory. (It is often referred to as the assumption of homoscedasticity.) When the error variance is not constant over all the observations, the error is said to be heteroscedastic. Heteroscedasticity can be removed by means of a suitable transformation. We describe an approach for detecting heteroscedasticity, its effects on the analysis, and finally, transformations for removing it from the data analyzed.

The dependent variable y, in a regression problem, may follow a probability distribution whose variance is a function of the mean of that distribution. The binomial and Poisson are two common probability distributions which have this characteristic. We know for example, that a variable that is distributed binomially with parameters n and π, has mean $n\pi$ and variance $n\pi(1 - \pi)$. It is also known that the mean and variance of a Poisson variable are equal. When the relationship between the mean and variance of a random variable is known, it is possible to find a simple transformation of the variable, which makes the variance approximately constant (stabilizes the variance). We list in Table 2.9, for convenience and easy reference, transformations that stabilize the variance for random variables with commonly occurring probability laws whose variances are functions of their means. See Bartlett (1947) or Kendall and Stuart (1968) pages 88–92, for a more detailed discussion.

The transformations listed in Table 2.9 not only stabilize the variance, but also have the effect of making the distribution of the transformed variable closer to the normal distribution. Consequently, these transformations

Table 2.9. **Transformations to stabilize variance**

Probability distribution of variable y	Variance of y in terms of its mean μ	Transformation	Resulting variance
Poisson[a]	μ	$\sqrt{y}$ or $(\sqrt{y} + \sqrt{y+1})$	0.25
Binomial[b]	$\dfrac{\mu(1-\mu)}{n}$	$\sin^{-1}\sqrt{y}$ (degrees)	$\dfrac{821}{n}$
		$\sin^{-1}\sqrt{y}$ (radians)	$\dfrac{0.25}{n}$
Negative binomial	$\mu + \lambda^2 m^2$	$\lambda^{-1}\sinh^{-1}(\lambda\sqrt{y})$ or	
		$\lambda^{-1}\sinh^{-1}(\lambda\sqrt{y} + 0.5)$	$\_0.25$

[a] For small values of y, $\sqrt{y + 0.5}$ is sometimes recommended.

[b] n is an index describing the size of sample; for $y = r/n$ a slightly better transformation is

$$\sin^{-1}\left(\frac{r+3/8}{n+3/4}\right)^{1/2}$$

serve the dual purpose of normalizing the variable as well as making the variance functionally independent of the mean.

As an illustration, consider the following situation: Let y be the number of accidents, and x the speed of operating a lathe in a machine shop. We want to study the relationship between the number of accidents y and the speed of lathe operation x. Suppose a linear relationship is postulated between y and x, and is given by

$$y = \beta_0 + \beta_1 x + u,$$

where u is the random error. The mean of y is seen to increase with x. It is known from empirical observation that rare events (events with small probabilities of occurrence) often have a Poisson distribution; and let us assume that y has a Poisson distribution. Since the mean and variance of y are the same (from the assumed probability law of y), it follows that the variance of y will not be constant, and consequently the assumption of homoscedasticity will not hold; in fact, the variance of y will be proportional to x. It is known, however, that the square root of a Poisson variable ($\sqrt{y}$) has a variance independent of the mean and is approximately equal to 0.25. To ensure homoscedasticity we, therefore, regress $\sqrt{y}$ against x. Here the transformation is chosen to stabilize the variance, the specific form being suggested by the probability law of the dependent variable.

An analysis of data employing transformations suggested by probabilistic considerations is demonstrated in the following example.

2.10. INJURY INCIDENTS IN AIRLINES

The number of injury incidents and the proportion of total flights from New York for 9 major United States, airlines for a single year is given in Table 2.10 and plotted in Figure 2.10. Let F_i and y_i denote the total flights and the number of injury incidents for the ith airline that year. Then the proportion of total flights N_i made by the ith airline is

$$N_i = \frac{F_i}{\sum F_i}.$$

It is thought that all the airlines are equally safe and the injury incidents can be explained by the model

$$y_i = \alpha + \beta N_i + u_i,$$

where α and β are constants, and u_i is the random error.

The results of fitting the model is given in Table 2.11. The plot of residuals against N_i is given in Figure 2.11.

The residuals in Figure 2.11 are seen to increase with N_i, and consequently the assumption of homoscedasticity seems to be violated. This is not surprising, since the injury incidents may behave as a Poisson variable which has a variance proportional to its mean. To ensure the assumption of homoscedasticity, as pointed out earlier, for a Poisson variable we make the square root transformation. Instead of working with y we work with $\sqrt{y}$, a variate which has an approximate variance of 0.25, and is more normally distributed than the original variable.

Consequently, the model we fit is

$$\sqrt{y} = \alpha' + \beta' N_i + u_i. \tag{2.10}$$

The result of fitting (2.10) is given in Table 2.12. The residuals from Equation (2.10) when plotted against N_i are shown in Figure 2.12.

Table 2.10. Number of injury incidents and proportion of total flights

	Y	N
001	11	0.0950
002	7	0.1920
003	7	0.0750
004	19	0.2078
005	9	0.1382
006	4	0.0540
007	3	0.1292
008	1	0.0503
009	3	0.0629

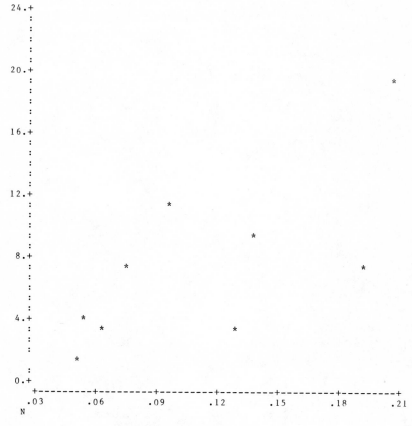

Fig. 2.10. Plot of *y* against *N*.

Table 2.11. Estimated regression coefficients (y versus N)

Variable	Coefficient	SE	t
N	64.975	25.196	2.58
CONSTANT	−0.140	3.141	−0.045
n = 9	s = 4.201	R² = 0.4872	

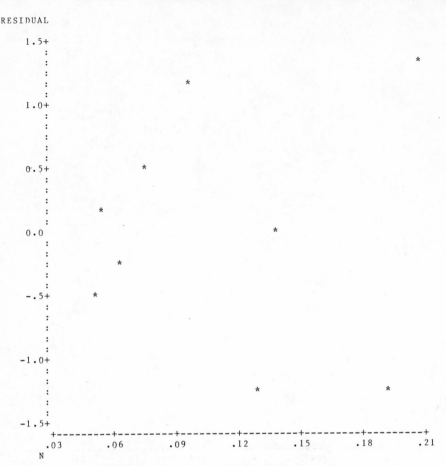

Fig. 2.11.   Plot of standardized residuals versus N.

The residuals for the transformed model do not seem to increase with N_i. This suggests that for the transformed model the homoscedastic assumption is not violated. The analysis of the model in terms of $\sqrt{y}$ and N can now proceed using standard techniques. The regression is significant here (as judged by the t statistic) but is not very strong. Only 48% of the total variability of the injury incidents of the airlines is explained by the variation in their number of flights. It appears that for a better explanation of injury incidents other factors have to be considered.

In the previous example the nature of the dependent variable (injury incidents) suggested that the error variance was not constant about the fitted line. The square root transformation was considered based on the

Table 2.12. *Estimated regression coefficients ($\sqrt{y}$ versus N)
(transformed model)*

Variable	Coefficient	SE	t
N	11.856	4.638	2.55
CONSTANT	1.169	0.058	2.02
$n = 9$	$s = 0.7733$	$R^2 = 0.4828$	

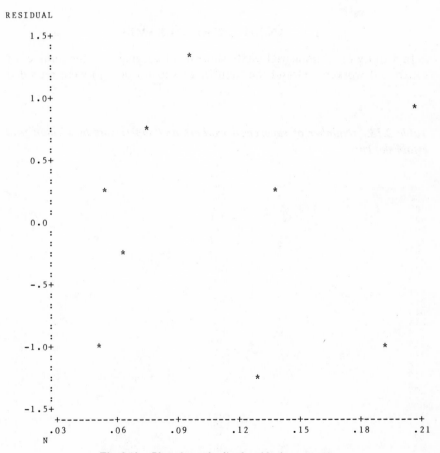

Fig. 2.12. Plot of standardized residuals versus N.

well-established empirical fact that the occurrence of accidents tend to follow the Poisson probability law. For Poisson variables, the square root is the appropriate transformation (Table 2.9). There are situations, however, when the error variance is not constant and there is no a priori reason to suspect that this would be the case. Empirical analysis will reveal the problem, and by making an appropriate transformation this effect can be eliminated. If the unequal error variance is not detected and eliminated, the resulting estimates will have large standard errors, but will be unbiased. This will have the effect of producing wide confidence intervals for the parameters, and tests with low sensitivity. We illustrate the method of analysis for a model with this type of heteroscedasticity in the next example.

2.11.   AN INDUSTRIAL EXAMPLE

In a study of 27 industrial establishments of varying size, the number of supervised workers (x) and the number of supervisors (y) were recorded

Table 2.13.   Number of supervised workers and supervisors in 27 industrial establishments

	X	Y
001	294	30
002	247	32
003	267	37
004	358	44
005	423	47
006	311	49
007	450	56
008	534	62
009	438	68
010	697	78
011	688	80
012	630	84
013	709	88
014	627	97
015	615	100
016	999	109
017	1022	114
018	1015	117
019	700	106
020	850	128
021	980	130
022	1025	160
023	1021	97
024	1200	180
025	1250	112
026	1500	210
027	1650	135

(Table 2.13). It was decided to study the relationship between the two variables, and as a start a linear model

$$y_i = \alpha + \beta x_i + u_i \qquad (2.11)$$

was postulated. A plot of y versus x suggests a simple linear model as a starting point (Figure 2.13). The results of fitting the linear model are given in Table 2.14.

Detection of Heteroscedastic Errors

The residual plots of e_i against x_i (Figure 2.14) shows that error variance tends to increase with x. The residuals tend to lie in a band which diverges

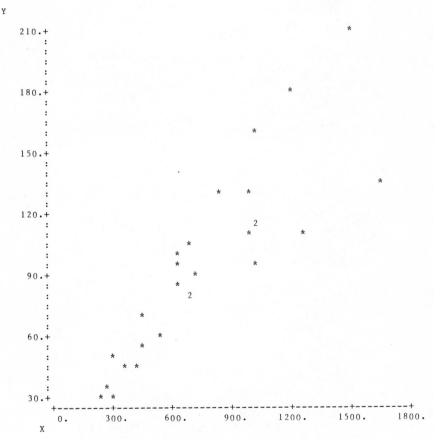

Fig. 2.13.   Number of Supervisors (y) versus number supervised (x).

Table 2.14. **Estimated regression coefficients**
(y versus x)

Variable	Coefficient	SE	t
x	0.115	.011	9.30
CONSTANT	14.448	9.562	1.51
$n = 27$	$s = 21.729$	$R^2 = 0.776$	

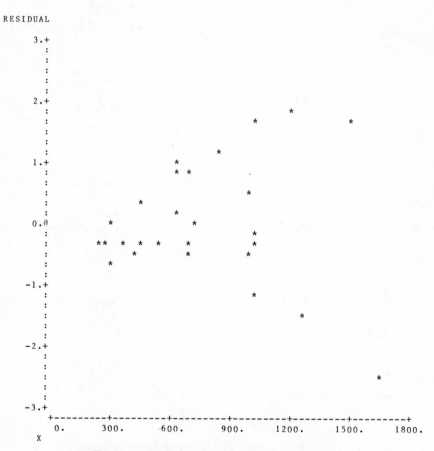

Fig. 2.14. Plot of standardized residuals against x.

as one moves along the x-axis. In general, if the bands within which the residuals lie diverges (i.e., becomes wider) as x increases, then the error variance is also increasing with x. On the other hand, if the bands converge (i.e., become narrower) the error variance decreases with x. If the bands that contain the residual plots are two lines parallel to the x-axis, then there is no evidence of heteroscedasticity. A plot of the standardized residuals against the independent variable points up the presence of heteroscedastic errors. In our present example, as can be seen from Figure (2.14) the residuals tend to increase with x.

2.12. REMOVAL OF HETEROSCEDASTICITY

In many industrial, economic, and biological applications, when unequal error variances are encountered, it is often found that the standard deviation of residuals tends to increase as the explanatory variable increases. Based on this empirical observation, we will hypothesize in the present example that the standard deviation of the residuals is proportional to x (some indication of this is available from the plot of the residuals in Figure 2.14).

$$\text{Var}(u_i) = k^2 x_i^2, \qquad k > 0. \tag{2.12}$$

Consider the model

$$\frac{y_i}{x_i} = \frac{\alpha}{x_i} + \beta + \frac{u_i}{x_i} \tag{2.13}$$

which is obtained by dividing both sides of (2.11) by x. Define a new set of variables and coefficients,

$$y' = \frac{y}{x}, \quad x' = \frac{1}{x}, \quad \alpha' = \beta, \quad \beta' = \alpha; \qquad u' = \frac{u}{x}.$$

In terms of the new variables (2.13) reduces to

$$y_i' = \alpha' + \beta' x_i' + u_i'. \tag{2.14}$$

Note that for the transformed model, variance of u_i' is constant and equals k^2. If our assumption about the error term as given in (2.12) holds, then to fit the model properly we must work with the transformed variables: y/x, $1/x$ as dependent and independent variables, respectively. If the fitted model for the transformed data is

$$\frac{y}{x} = a' + \frac{b'}{x} \tag{2.15}$$

then the fitted model in terms of the original variables is

$$y = b' + a'x. \tag{2.16}$$

The constant in the transformed model is the regression coefficient of x in the original model and vice versa. This can be seen from comparing (2.13) and (2.14).

The residuals obtained after fitting the transformed model are plotted in Figure 2.15. It is seen that the residuals are randomly distributed and lie roughly within two bands parallel to the $1/x$ axis which is the explanatory variable in the transformed model. There is no marked evidence of heteroscedasticity in the transformed model. The distribution of residuals shows no distinct pattern and we conclude that the transformed model is

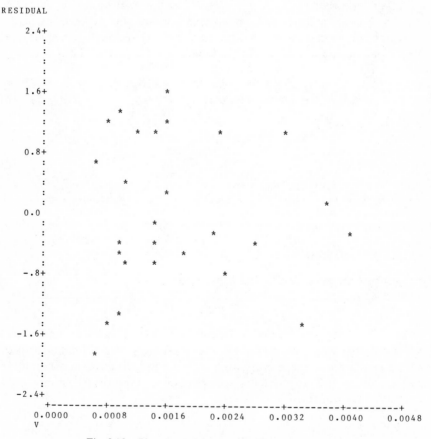

Fig. 2.15.   Plot of standardized residuals against $1/x$.

adequate. Our assumption about the error term appears to be correct; the transformed model has homoscedastic errors and the standard assumptions of least square theory hold. The result of fitting y/x and $1/x$ leads to estimates of a and b which can be used for the original model. The equation for the transformed variables is

$$\frac{y}{x} = .121 + \frac{3.803}{x}.$$

In terms of the original variables, we have

$$y = 3.803 + .121x.$$

The results are summarized in Table 2.15.

By comparing Tables 2.14 and 2.15 we see the reduction in standard errors which is accomplished by working with transformed variables. Variance of the estimate of the slope is reduced by 33%.

2.13. PRINCIPLE OF WEIGHTED LEAST SQUARES

Linear regression models with heteroscedastic errors can also be fitted by a method called the principle of weighted least squares (WLS). According to the principle, parameter estimates are obtained by minimizing a weighted sum of squares of residuals where the weights are inversely proportional to the variance of the errors. This is in contrast to ordinary least squares (OLS) where the parameter estimates are obtained by minimizing equally weighted sum of squares of residuals. In the previous example, the WLS estimates are obtained by minimizing

$$\sum \frac{1}{x_i^2}(y_i - \alpha - \beta x_i)^2 \tag{2.17}$$

as opposed to minimizing

$$\sum (y_i - \alpha - \beta x_i)^2. \tag{2.18}$$

Table 2.15. *Estimated regression coefficients of original equation when fitted by transformed variables*

Variable	Coefficient	SE	t
x	0.121	.009	13.44
CONSTANT	3.803	4.570	0.832
$n=27$	$s=22.577$	$R^2=0.7587$	

It can be shown that WLS is equivalent to performing OLS on the transformed variables y/x and $1/x$. We leave this as an exercise for the reader.

2.14. SUMMARY

After fitting a linear model one should examine the residuals for any evidence of heteroscedasticity. Heteroscedasticity is revealed if the residuals tend to increase or decrease with the values of the independent variable, and is conveniently examined from a plot of the residuals. If heteroscedasticity is present, account should be taken of this in fitting the model. If no account is taken of the unequal error variance, the resulting least square estimates will not have the maximum precision (smallest variances). Heteroscedasticity can be removed by working with transformed variables. Parameter estimates from the transformed model are then substituted for the appropriate parameters in the original model. The residuals from the appropriately transformed model should show no evidence of heteroscedasticity.

REFERENCES

Aitchison, J. and J. A. C. Brown, *The Lognormal Distribution,* Cambridge University Press, Cambridge, 1957.

Anscombe, F. J., Rejection of outliers, *Technometrics,* **2**, 123–167 (1960).

Anscombe, F. J. and J. W. Tukey, The examination and analysis of residuals, *Technometrics,* **5**, 141–160 (1963).

Bartlett, M. S., The use of transformations, *Biometrics,* **3**, 39–52 (1947).

Daniel, C. and F. S. Wood, *Fitting Equations to Data,* Wiley, New York, 1971.

Ellenberg, J. H., Testing for a single outlier from a general regression, *Biometrics,* **32**, 637–645 (1976).

Gallant, A. R., Nonlinear Regression, *American Statistician,* **29**, 73–81 (1975).

Goldberger, A. S., On the interpretation and estimation of Cobb–Douglas functions, *Econometrica,* **36**, 464–472 (1968).

Kendall, M. G. and A. Stuart, *The Advanced Theory of Statistics,* Vol. 3, Charles Griffin, London, 1968.

Rosner, B., On the detection of many outliers, *Technometrics,* **17**, 221–229 (1975).

CHAPTER 3

Multiple Regression Model

In this chapter the general multiple regression model is presented. The results are essentially extensions of the results given in Chapter 1 for the simple regression model. The presentation serves as a review of the standard results on regression analysis and is followed by an example which illustrates the application of the results. For proofs and derivations of these results the reader is referred to one of the texts on regression analysis listed at the end of the chapter.

3.1. DESCRIPTION OF THE DATA AND MODEL

The data consists of n observations on a dependent or response variable y and p independent (explanatory) variables $x_1, x_2, \ldots, x_p$. The observations are usually represented as follows:

Observation number	y	x_1	x_2	x_3	$\ldots$	x_p
1	y_1	x_{11}	x_{21}	x_{31}		x_{p1}
2	y_2	x_{12}	x_{22}	x_{32}		x_{p2}
3	y_3	x_{13}	x_{23}	x_{33}		x_{p3}
.	.	.	.	.		.
.	.	.	.	.		.
.	.	.	.	.		.
n	y_n	x_{1n}	x_{2n}	x_{3n}		x_{pn}

The relationship between y and $x_1, x_2, \ldots, x_p$ is formulated as a linear model

$$y_i = \beta_0 + \beta_1 x_{1i} + \beta_2 x_{2i} + \cdots + \beta_p x_{pi} + u_i, \qquad (3.1)$$

where $\beta_0, \beta_1, \beta_2, \ldots, \beta_p$ are constants referred to as the model partial regression coefficients (or simply as the regression coefficients) and u_i is a random disturbance. It is assumed that for any set of fixed values of $x_1, x_2, \ldots, x_p$ that fall within the range of the data, the linear Equation (3.1) provides an acceptable approximation to the true relationship between y

51

and the x's. In other words, y is approximately a linear function of the x's, and u_i measures the discrepancy in that approximation for the ith observation. In particular the u's contain no systematic information for determining y that is not already captured in the x's. It is assumed that u's are random quantities, independently distributed with zero means and constant variance σ^2.

The regression coefficient β_i may be interpreted as the increment in y corresponding to a unit increase in x_i when all other variables are held constant. Clearly this interpretation holds independently of the actual values of the x's.

The β's are estimated by minimizing the sum of squared residuals which is known as the method of least squares. Formally, it involves minimizing

$$S(\beta_0, \beta_1, \beta_2, \ldots, \beta_p) = \sum_{i=1}^{n} u_i^2$$

$$= \sum_{i=1}^{n} (y_i - \beta_0 - \beta_1 x_{1i} - \beta_2 x_{2i} \cdots - \beta_p x_{pi})^2.$$

By a direct application of calculus, it can be shown that the least square estimates $b_0, b_1, b_2, \ldots, b_p$, which minimize $S(\beta_0, \beta_1, \beta_2, \ldots, \beta_p)$ are given by the solution of the following system of equations.

$$S_{11}b_1 + S_{12}b_2 + S_{13}b_3 \cdots + S_{1p}b_p = S_{y1}$$
$$S_{12}b_1 + S_{22}b_2 + S_{23}b_3 \cdots + S_{2p}b_p = S_{y2}$$
$$\vdots$$
$$S_{1p}b_1 + S_{2p}b_2 + S_{3p}b_3 \cdots + S_{pp}b_p = S_{yp},$$

where

$$S_{ij} = \sum_{k=1}^{n} (x_{ik} - \bar{x}_i)(x_{jk} - \bar{x}_j), \qquad i,j = 1, 2, \ldots, p;$$

$$S_{yi} = \sum_{k=1}^{n} (y_k - \bar{y})(x_{ik} - \bar{x}_i), \qquad i = 1, 2, \ldots, p;$$

$$\bar{x}_i = \frac{\sum_{k=1}^{n} x_{ik}}{n}, \qquad \bar{y} = \frac{\sum_{k=1}^{n} y_k}{n},$$

and

$$b_0 = \bar{y} - b_1 \bar{x}_1 - b_2 \bar{x}_2 - b_3 \bar{x}_3 - \cdots - b_p \bar{x}_p.$$

The above system of equations are called the normal equations. b_0 is usually referred to as the intercept, and b_i as the estimate of the (partial) regression coefficient of x_i.

We assume that the system of equations is solvable and has a unique solution. We shall not say anything more about the actual process of solving the normal equations. We assume the availability of computer software that gives numerically accurate solutions.

Using the estimated regression coefficients we define a fitted (or predicted) value

$$\hat{y}_i = b_0 + b_1 x_{1i} + b_2 x_{2i} + \cdots + b_p x_{pi} \tag{3.2}$$

and an observed residual,

$$e_i = y_i - \hat{y}_i, \tag{3.3}$$

for each observation. The e_i's are used as before to evaluate model specification.

Properties of the least squares estimators are given in the following section. A reader familiar with matrix algebra will find a concise statement of the following results employing matrix notation in the Appendix.

3.2.   PROPERTIES OF THE LEAST SQUARE ESTIMATORS

1. b_i is an unbiased estimate of β_i and has a variance of $\sigma^2 c_{ii}$, where c_{ii} is the ith diagonal element of the inverse of corrected sums of squares and products matrix. The covariance between b_i and b_j is $\sigma^2 c_{ij}$, where c_{ij} is the element in the ith row and jth column of the inverse of the corrected sums of squares and products matrix. For all unbiased estimates which are linear in the observations the least square estimates have the smallest variance.

2.   An unbiased estimate of σ^2 is given by

$$s^2 = \frac{\text{SSE}}{n-p-1}, \tag{3.4}$$

where

$$\text{SSE} = \sum_{i=1}^{n} (y_i - b_0 - b_1 x_{1i} - b_2 x_{2i} \cdots - b_p x_{pi})^2. \tag{3.5}$$

With the additional assumption that the u_i's are normally distributed,

3. b_i is normally distributed with mean β_i and variance $\sigma^2 c_{ii}$.

4. $W = \text{SSE}/\sigma^2$ has a χ^2 distribution with $(n-p-1)$ degrees of freedom. The set of b_i's and s^2 are distributed independently of each other.

5.   The vector $b = (b_1, b_2, \ldots, b_p)$ has a p-variate normal distribution with mean vector $\beta = (\beta_1, \beta_2, \ldots, \beta_p)$ and variance–covariance matrix $\sigma^2(c_{ij})$.

The above results enable us to test various hypotheses about individual β's, and construct confidence intervals.

The statistic for testing $H_0(\beta_i = \beta_i^0)$, where β_i^0 is a constant chosen by the investigator, is

$$t = \frac{b_i - \beta_i^0}{s\sqrt{c_{ii}}} \qquad (3.6)$$

which has a Student's t distribution with $(n-p-1)$ degrees of freedom. The test is carried out by comparing the observed value with the appropriate critical t value. The usual test is for $\beta_i^0 = 0$ in which case t reduces to the ratio of b_i to its standard error. The confidence limits for β_i with confidence coefficient α are given by $b_i \pm t(n-p-1, \alpha/2)s\sqrt{c_{ii}}$, where $t(n-p-1, \alpha)$ is the $(1-\alpha)$ percentile point of the t distribution with $(n-p-1)$ degrees of freedom.

3.3.   PREDICTED VALUES AND STANDARD ERRORS

The fitted multiple regression equation can be used to predict the value of the dependent variable corresponding to an observation $x_0 = (x_{10}, x_{20}, x_{30}, \ldots, x_{p0})$. The predicted value $\hat{y}_0$ is given by

$$\hat{y}_0 = b_0 + b_1 x_{10} + b_2 x_{20} + \cdots + b_p x_{p0} \qquad (3.7)$$

and its variance is

$$\text{Var}(\hat{y}_0) = \sigma^2 \left(1 + \frac{1}{n}\right) + \sum_1^p (x_{i0} - \bar{x}_i)^2 V(b_i)$$

$$+ 2\sum_i \sum_{i<j} (x_{i0} - \bar{x}_i)(x_{j0} - \bar{x}_j)\text{Cov}(b_i, b_j), \qquad (3.8)$$

where

$$V(b_i) = \text{variance of } b_i,$$

$$\text{Cov}(b_i, b_j) = \text{covariance between } b_i \text{ and } b_j.$$

Confidence limits for $\hat{y}_0$ with confidence coefficient α are

$$\hat{y}_0 \pm t(n-p-1, \alpha/2) \text{s.e.}(\hat{y}_0),$$

where s.e.$(\hat{y}_0)$ is the standard error of $\hat{y}_0$ and is obtained by taking the square root of (3.8) after replacing σ^2 by s^2, where s^2 is the variance of the residuals and is given by (3.4).

As has already been mentioned in connection with simple regression, instead of predicting the response y corresponding to an observation $(x_{10}, x_{20}, x_{30}, \ldots, x_{p0})$, it may be desirable to predict the mean or the expected response corresponding to that observation. Let us denote the mean response at $(x_{10}, x_{20}, \ldots, x_{p0})$ by μ_0 and its estimate by $\hat{\mu}_0$. Then

$$\hat{\mu}_0 = b_0 + b_1 x_{10} + b_2 x_{20} + \cdots + b_p x_{p0}$$

as in Equation (3.7), but the variance of the mean predicted response for this observation is

$$\text{Var}(\hat{\mu}_0) = \frac{\sigma^2}{n} + \sum (x_{i0} - \bar{x}_i)^2 V(b_i)$$

$$+ 2 \sum_i \sum_{i<j} (x_{i0} - \bar{x}_i)(x_{j0} - \bar{x}_j) \text{Cov}(b_i, b_j), \qquad (3.9)$$

where $V(b_i)$ and $\text{Cov}(b_i, b_j)$ have the same meanings as before. The standard error of the mean predicted response corresponding to the observation $(x_{10}, x_{20}, \ldots, x_{p0})$ is obtained by taking the square root of (3.9) after replacing σ^2 by s^2.

3.4. MULTIPLE CORRELATION COEFFICIENT

After fitting the linear model to a given body of data, an assessment is made of the adequacy of fit. The most widely used measure is the multiple correlation coefficient R, or more frequently the square of the multiple correlation coefficient R^2.* There are several equivalent ways in which R^2 can be defined and interpreted. We define R^2 as

$$R^2 = 1 - \frac{\sum (y_i - \hat{y}_i)^2}{\sum (y_i - \bar{y})^2} \qquad (3.10)$$

---

*This quantity has already been introduced in Chapter 1. When there is only one independent variable, R is called the simple correlation coefficient.

and interpret it as the proportion of total variability which is explained by the regression equation. R^2 has a range between 0 and 1. When the model fits the data well, it is clear that the value of R^2 is close to unity. With a good fit, the observed and predicted values will be close to each other, and $\Sigma(y_i - \hat{y}_i)^2$ will be small. Then R^2 will be near to unity. On the other hand, if there is no relationship between the independent variable and the dependent variables and the linear model gives a poor fit, the best predicted value for an observation y_i would be $\bar{y}$; that is, in the absence of any relationship, the best estimate is the sample mean, for in that case the sample mean minimizes the sum of squared deviations. So in the absence of any linear relationship, R^2 will be near zero. The value of R^2 is therefore used as a *summary measure* to judge the fit of the linear model to a given body of data. As has already been pointed out in Chapter 1, a large value of R^2 does not necessarily mean that the data has been well fitted by the model. A more detailed analysis, as we outline in the next section, is needed to ensure that the model adequately describes the data.

3.5.   TESTS OF HYPOTHESES IN THE LINEAR MODEL

In addition to looking at hypotheses about individual β's several different hypotheses are considered in connection with the analysis of linear models. The most commonly investigated hypotheses are:

1.   All regression coefficients are zero. This implies that there is no linear relationship between the dependent variable and the set of independent variables.
2.   A subset of the regression coefficients are zero.
3.   A subset of the regression coefficients are equal to each other.

The different hypotheses about the regression coefficients can all be tested in the same way by a unified approach. Rather than describing the individual tests, we describe the general approach, and illustrate specific tests in the example that follows this section.

The model given in (3.1) will be referred to as the full model (FM). The null hypothesis to be tested specifies values for some of the regression coefficients. When these values are substituted in the full model, the resulting model is called the reduced model (RM). The number of *distinct* parameters to be estimated in the reduced model is smaller than the number of parameters to be estimated in the full model. The test for any hypothesis involves a comparison of the goodness of fit that is obtained when using the full model with the data, to the goodness of fit that results using the reduced model specified by the null hypothesis. If the reduced model gives as good a fit as the full model, then the null hypothesis, which defines the reduced model (by specifying some values of β_i), is not rejected. This procedure is formally described as follows.

Let $\hat{y}_i$ and $\hat{y}_i^*$ be the values predicted for y_i by the full model and the reduced model, respectively. The lack of fit in the data associated with the full model is $\Sigma(y_i - \hat{y}_i)^2$. We denote this by SSE(FM), the sum of squares due to error associated with the full model,

$$SSE(FM) = \sum (y_i - \hat{y}_i)^2. \tag{3.11}$$

The lack of fit in the data which is associated with the reduced model is $\Sigma(y_i - \hat{y}_i^*)^2$. This quantity is denoted by SSE(RM), the sum of squares due to error associated with the reduced model,

$$SSE(RM) = \sum (y_i - \hat{y}_i^*)^2. \tag{3.12}$$

In the full model there are $(p + 1)$ parameters $(\beta_0, \beta_1, \beta_2, \ldots, \beta_p)$ and let us suppose that for the reduced model there are k distinct parameters. To see whether the reduced model is adequate, we compare SSE(RM) − SSE(FM) with SSE(FM). Instead of comparing these two quantities directly, we use

$$\frac{SSE(RM) - SSE(FM)}{p + 1 - k} \quad \text{and} \quad \frac{SSE(FM)}{n - p - 1}.$$

The divisors are introduced to compensate for the different number of parameters which are involved in the two models as well as to ensure that the resulting test statistic has a standard statistical distribution. The ratio

$$F = \frac{[SSE(RM) - SSE(FM)]/(p + 1 - k)}{SSE(FM)/(n - p - 1)} \tag{3.13}$$

has the F distribution with $(p + 1 - k)$ and $(n - p - 1)$ degrees of freedom (d.f.). If the observed F value is large, in comparison to the tabulated value of F with $(p + 1 - k)$ and $(n - p - 1)$ d.f. at the 100α percent level, the result is significant at level α; that is, the reduced model is unsatisfactory and the null hypothesis, with its suggested values of β_i's in the full model, is rejected. The reader interested in proofs of the above statements is referred to Rao (1973), Searle (1971), Plackett (1960), or Seber (1977).

The distribution theory, confidence intervals, and tests of hypotheses outlined previously are valid only if the standard assumptions for the model residual terms hold. We have assumed that the residual terms are independent and normally distributed with mean zero and constant variance. When these assumptions are violated the standard results quoted previously do not hold and an application of them may lead to serious errors. As in the case of simple regression, graphical analysis of the observed residuals may be used to look for violations of the assumptions. There are more plots that may be examined in the multiple regression problem than in the case of simple regression. These plots will be described as they arise in the examples throughout the book.

Graphical analysis of the residuals may point to serious violations in one or more of the standard assumptions. These violations could invalidate the formal statistical inference procedures described above. Of more importance, the analysis of residuals may lead to suggestions of structure or point to information in the data that might be missed or overlooked if only summary statistics were evaluated. These suggestions or cues can lead to a better understanding and possibly a better model of the process under study. A careful graphical analysis of residuals may often prove to be the most important part of the regression analysis.

3.6. ASSUMPTIONS ABOUT THE EXPLANATORY VARIABLES

There are two assumptions concerning the explanatory variables that have not yet been discussed. These assumptions are more subtle than the earlier-mentioned standard least squares assumptions concerning the form of the equation and the probabilistic properties of the residual terms. The assumptions are (i) the explanatory variables are nonstochastic, that is, the values of the x's are fixed or selected in advance, and (ii) the x's are measured without error. These assumptions cannot be validated so they do not play a major role in the analysis. However, they do influence interpretation of the regression results.

The first assumption is satisfied only when the experimenter can set the values of x variables at predetermined levels. It is clear that under nonexperimental or observational situations this assumption will not be satisfied. The theoretical results that we have presented will still continue to hold, but their interpretation has to be modified. When the x's are not a fixed set of constants, all inferences are conditional, conditioned on the observed data. It should be noted that this conditional aspect of the inference is consistent with the approach to data analysis presented in this book. Our main objective is to extract the maximum amount of information from the available data.

The second assumption, that the independent variables are measured without error, is hardly ever satisfied. The errors in measurement will affect the residual variance, the multiple correlation coefficient, and also the individual estimates of the regression coefficients. The exact magnitude of the effects will depend on several factors, the most important of which are the standard deviation of the errors of measurement, and the correlation structure between the errors. The effect of the measurement errors will be to increase the residual variance, and reduce the magnitude of the observed multiple correlation coefficient. The effect of measurement errors on individual regression coefficients are more difficult to assess. The estimate of the regression coefficient for a variable is not only affected by its own measurement errors, but also by the measurement errors of other variables included in the equation. Correction for measurement errors on the estimated regression coefficients, even in the simplest case where all the measurement errors are uncorrelated, requires a knowledge of the ratio between the variances of the measurement errors for the several variables and the variance of the random error. Since these quantities are seldom, if ever, known (particularly in the social sciences, where this problem is most acute), we can never completely hope to remove the effect of measurement errors from the estimated regression coefficients. If the measurement errors are not large compared to the random errors, the effect of measurement errors is slight. In interpreting the coefficients in such an analysis, this point should be remembered. Although there is some problem in the estimation of the regression coefficients when the variables are in error, the regression equation may still be used for prediction. However, the presence of errors in the x variables decreases the accuracy of prediction. For a more extensive discussion of this problem the reader is referred to Cochran (1970).

The remainder of the chapter is devoted to an example that is used to illustrate some of the standard regression results. The example is based on data from a study in industrial psychology (management).

3.7.   A STUDY OF SUPERVISOR PERFORMANCE

Description of Data and Analysis

A recent survey of the clerical employees of a large financial organization included questions related to employee satisfaction with their supervisors. There was a question designed to measure the overall performance of a supervisor, as well as questions that related to specific activities involving interaction between supervisor and employee. An exploratory study was undertaken to try to explain the relationship between specific supervisor characteristics and overall satisfaction with supervisors as perceived by the employees. Initially, six questionnaire items were chosen as possible explanatory variables. Table 3.1 gives the description of the

Table 3.1.　*Description of variables in attitude survey*

Variable	Description
Y	Overall rating of job being done by supervisor
X_1	Handles employee complaints
X_2	Does not allow special privileges
X_3	Opportunity to learn new things
X_4	Raises based on performances
X_5	Too critical of poor performances
X_6	Rate of advancing to better jobs

variables in the study. As can be seen from the list, there are two broad types of variables included in the study. Variables X_1, X_2, and X_5 relate to direct interpersonal relationships between employee and supervisor, whereas variables X_3 and X_4 are of a less personal nature and relate to the job as a whole. Variable X_6 is not a direct evaluation of the supervisor but serves more as a general measure of how the employee perceives his or her own progress in the company. The data for the analysis was generated from the individual employee responses to the items on the survey questionnaire. The response on any item ranged from 1 through 5 indicating very satisfactory to very unsatisfactory, respectively. A dichotomous index was created for each item by collapsing the response scale to two categories {1,2} to be interpreted as a favorable response and {3,4,5} representing an unfavorable response. The data was collected in 30 departments selected at random from the organization. Each department had approximately 35 employees and one supervisor. The data to be used in the analysis is given in Table 3.2, and was obtained by aggregating responses for departments to get the proportion of favorable responses for each item for each department. The resulting data therefore consists of 30 observations on seven variables, one observation from each department.

We want to fit the model

$$Y = \beta_0 + \beta_1 X_1 + \beta_2 X_2 + \cdots + \beta_6 X_6 + u. \qquad (3.14)$$

The results of fitting the least square equation connecting Y and the six explanatory variables is given in Table 3.3. Before proceeding with the analysis we look at various plots of residuals to determine if there are any serious violations of model assumptions, or some model misspecification. We start by plotting the standardized residuals against the fitted values (Figure 3.1). There appears to be no systematic pattern of variation to the residuals. There are no outliers and all the standardized residuals lie within ± 2. The standardized residuals are next plotted against the various independent variables. For illustration we present in Figure 3.2 a plot of

Table 3.2. Attitude survey data

ROW	Y	X 1	X 2	X 3	X 4	X 5	X 6
* 1 *	43	51	30	39	61	92	45
* 2 *	63	64	51	54	63	73	47
* 3 *	71	70	68	69	76	86	48
* 4 *	61	63	45	47	54	84	35
* 5 *	81	78	56	66	71	83	47
* 6 *	43	55	49	44	54	49	34
* 7 *	58	67	42	56	66	68	35
* 8 *	71	75	50	55	70	66	41
* 9 *	72	82	72	67	71	83	31
* 10 *	67	61	45	47	62	80	41
* 11 *	64	53	53	58	58	67	34
* 12 *	67	60	47	39	59	74	41
* 13 *	69	62	57	42	55	63	25
* 14 *	68	83	83	45	59	77	35
* 15 *	77	77	54	72	79	77	46
* 16 *	81	90	50	72	60	54	36
* 17 *	74	85	64	69	79	79	63
* 18 *	65	60	65	75	55	80	60
* 19 *	65	70	46	57	75	85	46
* 20 *	50	58	68	54	64	78	52
* 21 *	50	40	33	34	43	64	33
* 22 *	64	61	52	62	66	80	41
* 23 *	53	66	52	50	63	80	37
* 24 *	40	37	42	58	50	57	49
* 25 *	63	54	42	48	66	75	33
* 26 *	66	77	66	63	88	76	72
* 27 *	78	75	58	74	80	78	49
* 28 *	48	57	44	45	51	83	38
* 29 *	85	85	71	71	77	74	55
* 30 *	82	82	39	59	64	78	39

Table 3.3. Regression coefficients, standard errors, and t values

Variable	Coefficient	SE	t
X_1	0.613	0.1610	3.81
X_2	−0.073	0.1357	−0.54
X_3	0.320	0.1685	1.90
X_4	0.081	0.2215	0.37
X_5	0.038	0.1470	0.26
X_6	−0.217	0.1782	−1.22
CONSTANT	10.787	11.5890	0.93
$n = 30$	$R^2 = 0.7326$	$s = 7.068$	

standardized residuals against X_1. None of the residuals give any evidence of gross violation of model assumptions or misspecification of the model. If the model were not specified correctly, and suppose for example, a term involving X_1^2 were needed, then the plot of residuals against X_1 would show systematic curvature as opposed to the random scatter that is typical of good model specification. In our present example, the standardized

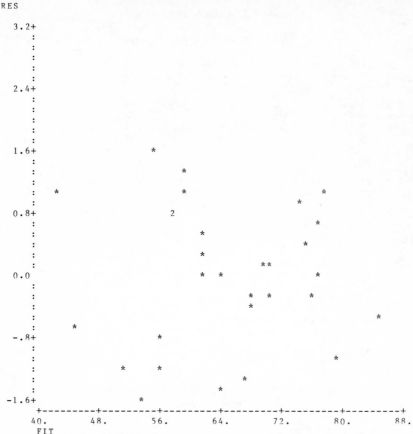

Fig. 3.1. Standardized residuals versus fitted values (attitude survey data).

residuals when plotted against X_i's $(i = 1, 2, \ldots, 6)$ show no trends, and consequently no evidence for model misspecification. We can now proceed with the analysis. The printed t values test the null hypothesis $H_0(\beta_i = 0)$ against an alternative $H_1(\beta_i \neq 0)$. From Table 3.3 it is seen that only the variables X_1 and X_3 have regression coefficients that approach being significantly different from zero.

The value of R^2 is 0.7326 showing that about 73% of the total variation in the overall rating of the job being done by the supervisor can be explained by the six variables. Although the t-test for the regression coefficients has already indicated that some of the regression coefficients $(\beta_1$ and $\beta_3)$ are significantly different from zero, we will, for illustrative purposes, test the hypothesis that the variables under consideration

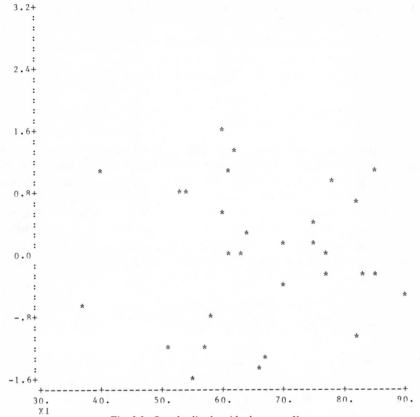

Fig. 3.2. Standardized residuals versus X_1.

$(X_1, X_2, \ldots, X_6)$ have no explanatory power, and all the regression coefficients are zero. The appropriate null hypothesis can be developed along the lines outlined in the previous section.

For the full model given in (3.14) we have to estimate seven parameters, six regression coefficients, and an intercept term β_0. The sum of squares due to error in the full model is

$$\text{SSE(FM)} = \sum (y_i - \hat{y}_i)^2 = 1149.$$

Under the null hypothesis, where all the β's are zero, the reduced model is

$$Y = \beta_0' + u. \qquad (3.15)$$

The number of parameters estimated for the reduced model is therefore 1. The estimate of β_0' as can be seen is $\bar{y}$, the mean of Y. Consequently,

$$\text{SSE(RM)} = \sum_1 (y_i - \bar{y})^2 = 4297.$$

The appropriate F is then found quite easily from (3.13), giving us

$$F = \frac{(4297 - 1149)/6}{1149/23} = 10.50, \qquad \text{d.f.} = 6, 23.$$

This test can also be expressed directly in terms of the sample multiple correlation coefficient. The null hypothesis which tests whether all the population regression coefficients are zero is equivalent to the hypothesis that states that the population multiple correlation coefficient is zero. Let R_p denote the sample multiple correlation coefficient which is obtained from fitting a model to n observations in which there are p independent (explaining) variables (i.e., we estimate p regression coefficients and one intercept). The appropriate F for testing $H_0(\beta_1 = \beta_2 = \cdots = \beta_p = 0)$ in terms of R_p is

$$F = \frac{R_p^2/p}{(1 - R_p^2)/(n-p-1)}, \qquad \text{d.f.} = p, n-p-1. \qquad (3.16)$$

In our present example the numerical equivalence is easily seen for

$$F = \frac{.7326/6}{.2674/23} = 10.50, \qquad \text{d.f.} = 6, 23.$$

The 1% F value for 6 and 23 degrees of freedom is 3.71, and since the observed F value is larger than this, the null hypothesis is rejected; all the β's cannot be taken as zero. This, of course, comes as no surprise, because of the large values of some t statistics. If any of the t tests for the individual regression coefficients prove significant, the F for testing all the regression coefficients zero will usually be significant. A more puzzling case can, however, arise when none of the t values for testing the regression coefficients are significant, but the F statistic given in (3.16) is significant. This implies that although none of the variables individually have significant explanatory power, the entire set of variables taken collectively explain a significant part of the variation in the dependent variable. This situation, when it occurs, should be looked at very carefully, for it may indicate a problem with the data analyzed, namely some of the explanatory variables may be highly correlated, a situation commonly called multicollinearity. We discuss this problem in Chapter 7.

3.8.   TESTING A SUBSET OF REGRESSION COEFFICIENTS EQUAL TO ZERO

We have so far attempted to explain Y in terms of six variables $X_1, X_2, X_3, \ldots, X_6$. We would like to examine whether Y can be explained adequately in terms of fewer variables. An important goal in regression analysis is to arrive at adequate descriptions of observed phenomenon in terms of as few meaningful variables as possible. This economy in description has two advantages. Firstly, it enables us to isolate the most important variables, and secondly, it provides us with a simpler description of the process studied, thereby making it easier to understand the process. Simplicity of description or the principle of parsimony as it is sometimes called is one of the important guiding principles in regression analysis.

To examine whether the variable Y can be explained in terms of fewer variables we look at a hypothesis that specifies that some of the regression coefficients are zero. If there are no overriding theoretical considerations as to which variables are to be included in the equation, the preliminary t-tests like those given in Table 3.3 are used to suggest the variables. In our current example suppose it was desired to explain the overall rating of the job being done by the supervisor by means of two variables, one taken from the group of personal employee-interaction variables X_1, X_2, X_5, and another taken from the group of variables X_3, X_4, X_6 which are of a less personal nature. From this point of view X_1 and X_3 suggest themselves, and we would like to examine if Y can be explained by X_1 and X_3 as adequately as the full set of six variables. The appropriate null hypothesis is

$$H_0(\beta_2 = \beta_4 = \beta_5 = \beta_6 = 0) \tag{3.17}$$

against the alternative that they are not all equal to zero. The reduced model in this case is

$$Y = \beta_0 + \beta_1 X_1 + \beta_3 X_3 + u. \tag{3.18}$$

The fitted least square equation is

$$Y = \begin{matrix} 9.871 \\ (7.062) \end{matrix} + \begin{matrix} 0.643 X_1 \\ (0.118) \end{matrix} + \begin{matrix} 0.211 X_3 \\ (0.134) \end{matrix} . \tag{3.19}$$

The quantities inside brackets directly under the coefficients are their respective standard errors. In the reduced model three parameters are estimated. The sum of squares due to error for the reduced model (3.18) is

$$SSE(RM) = \sum (y_i - \hat{y}_i^*)^2 = 1254.6,$$

where $\hat{y}^*$ are the values predicted by (3.19). From (3.13) it follows that the appropriate F is

$$F = \frac{(1254.6 - 1149)/4}{1149/23} = 0.528, \qquad \text{d.f.} = 4, 23.$$

The preceding test can also be expressed in terms of the sample multiple correlation coefficients. Let R_p denote the sample multiple correlation coefficient which is obtained when the full model with all the p variables in it is fitted to the data. Let R_q denote the sample multiple correlation coefficient when the model is fitted with q specific variables; that is, the null hypothesis states that $(p-q)$ specified variables have zero regression coefficient. The F statistic for testing the above hypothesis is

$$F = \frac{\left(R_p^2 - R_q^2\right)/(p-q)}{\left(1 - R_p^2\right)/(n-p-1)}, \qquad \text{d.f.} = p - q, n - p - 1. \qquad (3.20)$$

In our present example $n = 30$, $p = 6$, $q = 2$, $R_6^2 = .7326$, and $R_2^2 = .7080$. Substituting these in (3.20) we get an F value of 0.528 as before.

The value of F is not significant and the null hypothesis is not rejected. The variables X_1 and X_3 together explain the variation in Y as adequately as the full set of X's. At this stage residual plots similar to those described earlier are examined to see if the deletion of variables X_2, X_4, X_5, X_6 has caused any violations of the model assumptions. In our present example, the residual plots appear satisfactory, and we conclude that deletion of X_2, X_4, X_5, X_6 does not adversely affect the explanatory power of the model.

3.9. TESTING THE EQUALITY OF REGRESSION COEFFICIENTS

By the general method outlined in Section 3.5 it is possible to test the equality of two regression coefficients in the same model. In the present example we will test whether the regression coefficient of the variables X_1 and X_3 can be treated as equal. The test is performed assuming that it has already been established that the regression coefficients for X_2, X_4, X_5, and X_6 are zero. The null hypothesis to be tested is

$$H_0(\, \beta_1 = \beta_3 |\, \beta_2 = \beta_4 = \beta_5 = \beta_6 = 0). \qquad (3.21)$$

The full model assuming $\beta_2 = \beta_4 = \beta_5 = \beta_6 = 0$ is

$$Y = \beta_0 + \beta_1 X_1 + \beta_3 X_3 + u. \qquad (3.22)$$

Under the null hypothesis, the reduced model is

$$Y = \beta_0' + \beta'_1(X_1 + X_3) + u. \tag{3.23}$$

A simple way to carry out the test is to fit the model given by (3.22) to the data. The resulting equation and the multiple correlation coefficient for this equation has been given in the previous section.

We next fit the reduced model given in (3.23). This can be done quite simply by generating a new variable $W = X_1 + X_3$ and fitting the model

$$Y = \beta_0' + \beta_1' W + u. \tag{3.24}$$

The least squares estimates of β_0', β_1' and the sample multiple correlation coefficient (in this case it is the simple correlation coefficient between y and W since we have only two variables) is obtained. The fitted equation is

$$Y = 9.988 + 0.444 W$$

with a multiple correlation coefficient of 0.8176.

The appropriate F for testing the null hypothesis is

$$F = \frac{(.7080 - .6685)/1}{(1 - .7080)/27} = 3.65, \qquad \text{d.f.} = 1, 27.$$

The resulting F is not significant; the null hypothesis is not rejected. The distribution of the residuals for this equation (not given here) was found satisfactory.

The equation

$$Y = 9.988 + 0.444(X_1 + X_3)$$

is not inconsistent with the given data. We conclude then that each of X_1 and X_3 have the same incremental effect in determining employee satisfaction with a supervisor. This test could also be performed by using a t statistic, given by

$$t = \frac{b_1 - b_3}{\text{s.e.}(b_1 - b_3)}$$

with 27 degrees of freedom. The conclusions are identical and follow from the fact that F with 1 and p degrees of freedom is equal to the square of t with p d.f.

In this example we have discussed a sequential or step-by-step approach to model building. We have discussed the equality of β_1 and β_3 under the assumption that the other regression coefficients are equal to zero. We can, however, test a more complex null hypothesis which states that β_1 and β_3 are equal and $\beta_2, \beta_4, \beta_5, \beta_6$ are all equal to zero. This null hypothesis H_0' is formally stated as

$$H_0'(\beta_1 = \beta_3, \beta_2 = \beta_4 = \beta_5 = \beta_6 = 0). \tag{3.25}$$

The difference between (3.21) and (3.25) is that in (3.21) β_2, β_4, β_5 and β_6 are assumed zero whereas in (3.25) this is under test. The null hypothesis (3.25) can be tested quite easily. The reduced model under H_0' is (3.23) but this model is not compared to the model of equation (3.22) as in the case of H_0, but with the model given in (3.14), the full model with all six variables in the equation. The F statistic for testing H_0' therefore is

$$F = \frac{(.7326 - .6685)/5}{.2674/23}$$

$$= 1.10, \qquad \text{d.f.} = 5, 23.$$

The result is insignificant as before. The first test is more sensitive for detecting departures from equality of the regression coefficients than the second test.

3.10.   ESTIMATING AND TESTING OF REGRESSION PARAMETERS UNDER CONSTRAINTS

Sometimes in fitting regression equations to a given body of data it is desired to impose a constraint on the values of the parameters. A common constraint is that the regression coefficients sum to a specified value, usually unity. The constraints often arise because of some theoretical or physical relationships which may connect the variables. Although no such relationships are obvious in our present example, we consider $\beta_1 + \beta_3 = 1$ for the purpose of demonstration. Assuming that the model given in (3.22) has already been accepted, we may futher argue that if each of X_1 and X_3 are increased by a fixed amount, then Y should also increase by that same amount. Formally we are led to the null hypothesis H_0 which states

$$H_0(\beta_1 + \beta_3 = 1 | \beta_2 = \beta_4 = \beta_5 = \beta_6 = 0). \tag{3.26}$$

Under H_0 the reduced model is

$$Y = \beta_0 + \beta_1 X_1 + (1 - \beta_1) X_3 + u,$$

$$Y - X_3 = \beta_0 + \beta_1(X_1 - X_3) + u,$$

or

$$Y' = \beta_0 + \beta_1 V + u,$$

where

$$Y' = Y - X_3, \quad \text{and} \quad V = X_1 - X_3.$$

The least squares estimates of the parameters β_1 and β_3 under the constraint is obtained by fitting a regression equation with Y' as dependent variable and V as the independent variable. The fitted equation is

$$Y = 1.166 + 0.694 X_1 + 0.306 X_3$$

and the multiple correlation coefficient is 0.6905.

The test for H_0 is given by

$$F = \frac{(.7080 - .6905)/1}{.2920/27} = 1.62, \qquad \text{d.f.} = 1,27,$$

which is not significant. The data supports the proposition that the sum of the partial regression coefficients of X_1 and X_3 equal unity.

Recall that we have now tested two separate hypotheses about β_1 and β_3, one which states that they are equal and the other that they sum to unity. Since both of them hold, it is implied that both coefficients can be taken to be 0.5. A test of that null hypothesis, $\beta_1 = \beta_3 = 0.5$, may be performed directly by applying the methods we have outlined.

The previous example in which the equality of β_1 and β_3 was investigated can be considered as a special case of constrained problem, in which the constraint is $\beta_1 - \beta_3 = 0$. The tests for the full set or subsets of regression coefficients being zero can also be thought of as examples of testing regression coefficients under constraints.

From the above discussions it is clear that several models may describe a given body of data adequately. Where several descriptions of the data are available, it is important that they all be considered. Some descriptions may be more meaningful than others (meaningful being judged in the context of the application and considerations of subject matter), and one of them may be finally adopted. Looking at alternative descriptions of the data provides insight that might be overlooked in focusing on a single description.

The question of which variables to include in a regression equation is very complex and is taken up in detail in Chapter 9. We make two remarks here that will be elaborated on in later chapters.

1. The estimates of regression coefficients which do not significantly differ from zero are most commonly replaced by zero in the equation. The replacement has two advantages: a simpler model, and smaller prediction variance ($\text{Var}(\hat{y})$).

2. A variable or a set of variables may sometimes be retained in an equation because of their theoretical importance in a given problem, even though the sample regression coefficients are statistically insignificant. That is, sample coefficients which are not significantly different from zero are not replaced by zero. The variables so retained should give a meaningful process description, and the coefficients help to assess the contributions of the X's to the value of the dependent variable Y.

3.11. SUMMARY

We have illustrated the testing of various hypotheses in connection with the linear model. Rather than describing individual tests we have outlined a general procedure by which they can be performed. It has been shown that the various tests can all be described in terms of the appropriate sample multiple correlation coefficients. It is to be emphasized here, that before starting on any testing procedure, the adequacy of the model assumptions should always be examined. Residual plots provide a very convenient graphical way of accomplishing this task. The test procedures will not be valid if the assumptions on which the tests are based do not hold. If a new model is chosen on the basis of a statistical test, residuals from the new model should be examined before terminating the analysis. It is only by careful attention to detail that a satisfactory analysis of data can be carried out.

REFERENCES

Brownlee, K. A., *Statistical Theory and Methodology in Science and Engineering*, Wiley, New York, 1965.

Cochran, W. G., Some effects of errors of measurement on multiple correlation, *Amer. Stat. Assoc.*, **65**, 22–34 (1970).

Johnston, J., *Econometric Methods*, McGraw-Hill, New York, 1972.

Plackett, R. L., *Regression Analysis*, Oxford University Press, London, 1960.

Rao, C. R., *Linear Statistical Inference and its Applications*, Wiley, New York, 1973.

Searle, S. R., *Linear Models*, Wiley, New York, 1971.

Seber, G. F., *Linear Regression Analysis*, Wiley, New York, 1977.

APPENDIX

We present the standard results of multiple regression analysis in matrix notation. Let us define the following matrix and vectors:

$$
\mathbf{X} = \begin{bmatrix} x_{01} & x_{11} & \cdots & x_{p1} \\ x_{02} & x_{12} & \cdots & x_{p2} \\ \cdot & \cdot & & \cdot \\ \cdot & \cdot & & \cdot \\ \cdot & \cdot & & \cdot \\ x_{0n} & x_{1n} & \cdots & x_{pn} \end{bmatrix}, \quad \mathbf{y} = \begin{bmatrix} y_1 \\ y_2 \\ \cdot \\ \cdot \\ \cdot \\ y_n \end{bmatrix}, \quad \mathbf{u} = \begin{bmatrix} u_1 \\ u_2 \\ \cdot \\ \cdot \\ \cdot \\ u_n \end{bmatrix}
$$

and

$$
\boldsymbol{\beta} = \begin{bmatrix} \beta_0 \\ \beta_1 \\ \cdot \\ \cdot \\ \cdot \\ \beta_p \end{bmatrix}.
$$

The expressions in the appendix are numbered identically as those in the main text.

The linear model which represents the data is

$$
\mathbf{Y} = \mathbf{X}\boldsymbol{\beta} + \mathbf{u}, \tag{3.1$'$}
$$

where $x_{0i} = 1$ for all i.

The assumptions made about u for least square estimation are

$$
E(\mathbf{u}) = \mathbf{0}, \ \mathrm{Var}(\mathbf{u}) = E(\mathbf{uu}') = \sigma^2 \mathbf{I}_n;
$$

that is, u_i's are independent, have zero mean and constant variance. This implies

$$
E(\mathbf{y}) = \mathbf{X}\boldsymbol{\beta}
$$

The least squares estimator $\mathbf{b}$ of $\boldsymbol{\beta}$ is obtained by minimizing the sum of squared deviations of the observations from their expected values. Hence, the least squares estimators are obtained by minimizing $S(\boldsymbol{\beta})$, where

$$
S(\boldsymbol{\beta}) = \mathbf{u}'\mathbf{u} = (\mathbf{Y} - \mathbf{X}\boldsymbol{\beta})'(\mathbf{Y} - \mathbf{X}\boldsymbol{\beta}).
$$

Minimization of $S(\boldsymbol{\beta})$ leads to the system of equations

$$(\mathbf{X'X})\mathbf{b} = \mathbf{X'Y}.$$

This system of equations is called the normal equations. Assuming that $(\mathbf{X'X})$ has an inverse, $\mathbf{b}$ can be written explicitly as

$$\mathbf{b} = (\mathbf{X'X})^{-1}\mathbf{X'Y}.$$

The vector of predicted values $\hat{\mathbf{y}}$ corresponding to the observed $\mathbf{y}$ is

$$\hat{\mathbf{y}} = \mathbf{Xb}. \tag{3.2}'$$

The vector of residuals $\mathbf{e}$ is given by

$$\mathbf{e} = \mathbf{y} - \hat{\mathbf{y}} = \mathbf{y} - \mathbf{Xb}. \tag{3.3}'$$

The properties of the least squares estimators are

1. $\mathbf{b}$ is an unbiased estimator of $\boldsymbol{\beta}$, with variance-covariance matrix $\mathrm{Var}(\mathbf{b})$ which is

$$\mathrm{Var}(\mathbf{b}) = E(\mathbf{b}-\boldsymbol{\beta})(\mathbf{b}-\boldsymbol{\beta})' = \sigma^2(\mathbf{X'X})^{-1} = \sigma^2\mathbf{C},$$

where

$$\mathbf{C} = (\mathbf{X'X})^{-1} \quad \text{and} \quad E(\mathbf{b}) = \boldsymbol{\beta}.$$

Of all unbiased estimators of $\boldsymbol{\beta}$ which are linear in the observations, the least square estimator has minimum variance.

2. An unbiased estimator of σ^2 is s^2 where

$$s^2 = \frac{\mathbf{e'e}}{n-p-1} = \frac{(\mathbf{y}-\hat{\mathbf{y}})'(\mathbf{y}-\hat{\mathbf{y}})}{n-p-1} = \frac{\mathbf{y'y}-\mathbf{b'x'y}}{n-p-1}. \tag{3.4}'$$

With the added assumption that u_i's are normally distributed we have

3. The vector $\mathbf{b}$ has a p-variate normal distribution with mean vector $\boldsymbol{\beta}$ and variance–covariance matrix $\sigma^2\mathbf{C}$. The marginal distribution of b_i is normal with mean β_i and variance $\sigma^2 C_{ii}$, where C_{ii} is the ith diagonal element of $\mathbf{C}$.

4. The quantity $W = \mathbf{e'e}/\sigma^2$ has an χ^2 distribution with $(n-p-1)$ degrees of freedom.

5. $\mathbf{b}$ and s^2 are distributed independently of one another.

The predicted value $\hat{y}_0$ corresponding to an observation vector $\mathbf{x}_0' = (x_{00}, x_{10}, x_{20}, \ldots, x_{p0})$ with $x_{00} = 1$ is

$$\hat{y}_0 = \mathbf{x}_0' \mathbf{b} \tag{3.7}'$$

and its variance is

$$\mathrm{Var}(\hat{y}_0) = \left(\mathbf{x}_0' (\mathbf{X}'\mathbf{X})^{-1} \mathbf{x}_0 + 1 \right) \sigma^2. \tag{3.8}'$$

The mean response $\hat{\mu}_0$ corresponding to $\mathbf{x}_0'$ is

$$\hat{\mu}_0 = \mathbf{x}_0' \mathbf{b} \tag{3.9}'$$

with variance $V(\hat{\mu}_0) = \mathbf{x}_0' (\mathbf{X}'\mathbf{X})^{-1} \mathbf{x}_0 \sigma^2$.

CHAPTER 4

Qualitative Variables as Regressors

4.1. INTRODUCTION: INDICATOR VARIABLES

Qualitative factors can be very useful as explanatory variables in regression analysis. Factors such as sex, marital status, or political affiliation can be represented by indicator or dummy variables. These variables take on only two values, usually zero and one. The two values signify that the observation belongs in one of two possible categories. The numerical values of dummy variables are not intended to reflect a quantitative ordering of the categories, but only serve to identify category or class membership. For example, an analysis of salaries earned by computer programmers may include factors such as education, years of experience, and sex as explanatory variables. The sex variable could be quantified, say, as 1 for female and 0 for male. Dummy variables can also be used in a regression equation to distinguish among three or more groups as well as among classifications across various types of groups. For example the regression described above may also include a dummy variable to distinguish whether the observation was for a systems or applications programmer. The four conditions determined by sex and type of programming can be represented by combining the two variables.

Dummy variables can be used in a variety of ways and may be considered whenever there are qualitative factors affecting a relationship. We shall illustrate some of the applications with examples and suggest some additional applications. It is hoped that the reader will recognize the general applicability of the technique from the examples. In the first example, we look at data on a salary survey, such as the one mentioned above, and use dummy variables to adjust for various categorical factors that affect the regression relationship. The second example uses dummy variables for analyzing and testing for equality of regression relationships in various subsets of a population

4.2. SALARY SURVEY DATA

The first set of data was developed from a salary survey of computer professionals in a large corporation. The objective of the survey was to identify and quantify those factors that determine salary differentials. In addition, the data could be used to determine if the corporation's salary administration guidelines were being followed. The data appears in Table 4.1. The variables are (1) experience (EXPRNC) measured in years, (2) education (EDUC) coded as 1 for completion of high school, 2 for completion of a college, and 3 for the completion of an advanced degree, and (3) MGT which is coded as 1 for a person with management responsibility and 0 otherwise. We shall try to measure the effects of these three variables on salary using regression analysis.

A linear relationship will be used for salary and experience. We shall assume that each additional year of experience is worth a fixed salary increment. Education may also be treated in a linear fashion. If the education variable is used in the regression equation in raw form, we would be assuming that each step up in education is worth a fixed increment to salary. That is, with all other variables held constant, the relationship between salary and education is linear. That interpretation is possible, but may be too restrictive. Instead, we shall view education as a categorical variable and define two dummy variables to represent the three categories. These two variables allow us to pick up the effect of education on salary whether or not it is linear. The management variable is also a dummy variable designating the two categories, 1 for management positions and 0 for regular staff positions.

Note that when using dummy variables to represent a set of categories, the number of these variables required is one less than the number of categories. For example, in the case of the education categories above, we have

$$E_{1i} = \begin{array}{ll} 1 & \text{if the } i\text{th respondent is in the H.S. category} \\ 0 & \text{otherwise} \end{array}$$

and

$$E_{2i} = \begin{array}{ll} 1 & \text{if the } i\text{th respondent is in the B.S. category} \\ 0 & \text{otherwise.} \end{array}$$

As stated above, these two variables taken together uniquely represent the three groups. For H.S., $E_1 = 1$, $E_2 = 0$, for B.S., $E_1 = 0$, $E_2 = 1$ and for advanced degree, $E_1 = 0$, $E_2 = 0$. Furthermore, if there were a third vari-

Table 4.1. Salary survey data

	EXPRNC	EDUC	MGT	SALARY
001	1	1	1	13876
002	1	3	0	11608
003	1	3	1	18701
004	1	2	0	11283
005	1	3	0	11767
006	2	2	1	20872
007	2	2	0	11772
008	2	1	0	10535
009	2	3	0	12195
010	3	2	0	12313
011	3	1	1	14975
012	3	2	1	21371
013	3	3	1	19800
014	4	1	0	11417
015	4	3	1	20263
016	4	3	0	13231
017	4	2	0	12884
018	5	2	0	13245
019	5	3	0	13677
020	5	1	1	15965
021	6	1	0	12336
022	6	3	1	21352
023	6	2	0	13839
024	6	2	1	22884
025	7	1	1	16978
026	8	2	0	14803
027	8	1	1	17404
028	8	3	1	22184
029	8	1	0	13548
030	10	1	0	14467
031	10	2	0	15942
032	10	3	1	23174
033	10	2	1	23780
034	11	2	1	25410
035	11	1	0	14861
036	12	2	0	16882
037	12	3	1	24170
038	13	1	0	15990
039	13	2	1	26330
040	14	2	0	17949
041	15	3	1	25685
042	16	2	1	27837
043	16	2	0	18838
044	16	1	0	17483
045	17	2	0	19207
046	20	1	0	19346

Source: Created by the authors.

able, E_{3i} defined to be 1 or 0 depending on whether the ith respondent is in the advanced degree category or not, then for each respondent, $E_1 + E_2 + E_3 = 1$. Then $E_3 = 1 - E_1 - E_2$ clearly showing that one of the variables is superfluous.* Similarly there is only one dummy variable required to distinguish the two management categories.

*If E_1, E_2, and E_3 were used there would be a perfect linear relationship among the explanatory variables which is an extreme case of multicollinearity as described in Chapter 7.

In terms of the dummy variables described above, the regression model is

$$\text{SALARY} = \beta_0 + \beta_1 \text{EXPRNC} + \gamma_1 E_1 + \gamma_2 E_2 + \delta_1 \text{MGT} + u. \qquad (4.1)$$

By evaluating Equation (4.1) for the different values of the dummy variables, it follows that there is a different regression equation for each of the six (three education and two management) categories, namely:

EDUC	MGT	Regression equation		
H.S.	Mgt resp.	$\text{SALARY} = (\beta_0 + \gamma_1 + \delta_1)$	$+ \beta_1 \text{EXPRNC} + u$	
H.S.	None	$\text{SALARY} = (\beta_0 + \gamma_1)$	$+ \beta_1 \text{EXPRNC} + u$	
B.S.	Mgt resp.	$\text{SALARY} = (\beta_0 + \gamma_2 + \delta_1)$	$+ \beta_1 \text{EXPRNC} + u$	(4.2)
B.S.	None	$\text{SALARY} = (\beta_0 + \gamma_2)$	$+ \beta_1 \text{EXPRNC} + u$	
Adv.	Mgt resp.	$\text{SALARY} = (\beta_0 + \delta_1)$	$+ \beta_1 \text{EXPRNC} + u$	
Adv.	None	$\text{SALARY} = \beta_0$	$+ \beta_1 \text{EXPRNC} + u$	

According to the proposed model, we may say that the dummy variables help to determine the base salary level as a function of education and management status after adjustment for years of experience.

The results of the regression computations for the model given in Equation (4.1) appear in Table 4.2. The proportion of salary variation explained by the model is quite high, .9568. At this point in the analysis we should investigate the pattern of residuals to check on model specification. We shall postpone that investigation for now and assume that the model is satisfactory so that we can discuss the interpretation of the regression results. We shall return to analyze the residuals later and find that the model must be altered.

We see that the coefficient of EXPRNC is 546.16. That is, each additional year of experience is estimated to be worth an annual salary increment of $546.16. The other coefficients may be interpreted by looking

Table 4.2. Regression analysis of salary data

Variable	Coefficient	SE	t
EXPRNC	546.16	30.52	17.90
E_1	-2996.00	411.73	-7.28
E_2	147.98	387.64	0.38
MGT	6883.50	313.91	21.93
CONSTANT	11032.00	383.20	28.79
$n = 46$	$R^2 = .9568$	$s = 1027.39$	

into Equations (4.2). The coefficient of the management indicator variable, δ_1, is estimated to be 6883.50. From (4.2) we interpret this amount to be the average incremental value in annual salary associated with a management position. For the education variables, γ_1 measures the salary differential for the H.S. category relative to the advanced degree category and γ_2 measures the differential for the B.S. category relative to the advanced degree category. The difference, $\gamma_2 - \gamma_1$, measures the differential salary for the H.S. category relative to the B.S. category. From the regression results, in terms of salary for computer professionals, we see that an advanced degree is worth $2996 more than a high school diploma, a B.S. is worth $147.98 more than an advanced degree (this differential is not statistically significant, $t = .382$), and a B.S. is worth about $3144 more than a high school diploma. These salary differentials hold for every fixed level of experience.

Returning to the question of model specification, consider Figure 4.1 where the regression residuals are plotted against EXPRNC. The plot suggests that there may be three or more specific levels of residuals. Possibly the dummy variables that have been defined are not adequate for explaining the effects of education and management status. The residual plot is reconstructed in Figure 4.2 where each residual is identified with one of the six education–management combinations. We see that the residuals cluster by size according to their education–management category. Another way to observe this effect is to plot the residuals against a new categorical variable that takes a separate value for each of the six combinations. This graph is, in effect, a plot of residuals versus a potential explanatory variable that has not yet been used in the equation. The graph, appearing as Figure 4.3, suggests that the combinations of education and management have not been satisfactorily treated in the model. Within each of the six groups, the residuals are either almost totally positive or totally negative. This behavior implies that the model given in Equation (4.1) does not adequately explain the relationship between salary, experience, education, and management. The graph points to some hidden structure in the data that has not been explored.

The graphs strongly suggest that the effects of education and management status on salary determination are not additive. Note that in the model in Equation (4.1) and its further exposition in Equations (4.2), the incremental effects of both variables are determined by additive constants. For example, the effect of a management position is measured as δ_1, independently of the level of educational attainment. Nonadditive effects of these variables can be evaluated by constructing additional variables that are used to measure what may be referred to as multiplicative or interaction effects. These variables are defined as products of the existing

RESIDUAL

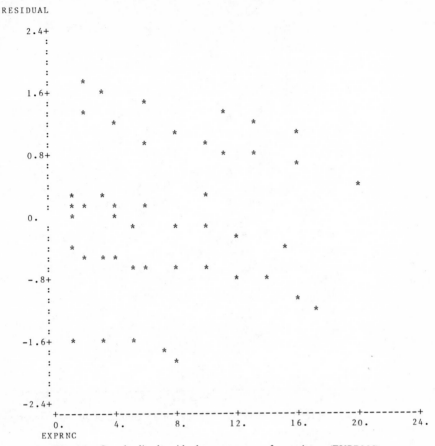

Fig. 4.1. Standardized residuals versus years of experience (EXPRNC).

dummy variables, namely $(E_1 \cdot \text{MGT})$ and $(E_2 \cdot \text{MGT})$. The inclusion of these two variables on the right-hand side of Equation (4.1) leads to a model that is no longer additive in education and management, but recognizes the multiplicative effect of these two variables.

The expanded model is

$$\text{SALARY} = \beta_0 + \beta_1 \text{EXPRNC} + \gamma_1 E_1 + \gamma_2 E_2 + \delta_1 \text{MGT}$$

$$+ \alpha_1 (E_1 \cdot \text{MGT}) + \alpha_2 (E_2 \cdot \text{MGT}) + u \qquad (4.3)$$

The regression results are in Table 4.3. The residuals from the regression on the expanded model are plotted against EXPRNC in Figure 4.4. Note

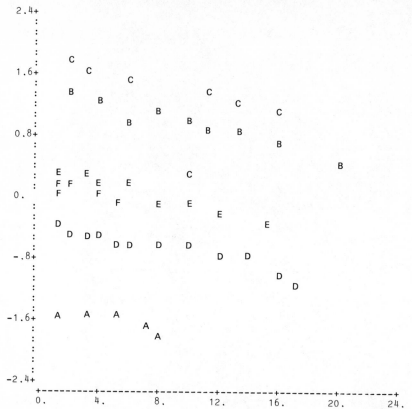

Fig. 4.2. Standardized residuals versus years of experience identified by education–management category.

that observation number 33 is an outlier. Salary is overpredicted by the model. Checking this observation in the listing of the raw data, it appears that this particular respondent seems to have fallen behind by a couple of hundred dollars in annual salary as compared to other respondents with similar characteristics. To be sure that this one observation is not overly affecting the regression estimates, it has been deleted and the regression rerun. The new results are in Table 4.4.

The regression coefficients are basically unchanged. However, the standard deviation of the residuals has been reduced to $67.28 and the proportion of explained variation has reached .9998. The plot of residuals versus EXPRNC (Figure 4.5) appears to be satisfactory as compared with the similar residual plot for the additive model. In addition, the plot of residuals for each education–management category (Figure 4.6) shows that

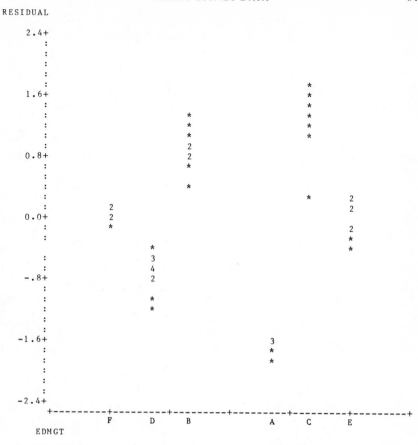

Fig. 4.3. Standardized residuals versus education – management categorical variable.

Table 4.3.   Regression analysis of salary data: Expanded model

Variable	Coefficient	SE	t
EXPRNC	496.97	5.57	89.22
E_1	− 1731.00	105.38	− 16.43
E_2	− 349.00	97.61	− 3.57
MGT	7047.30	102.63	68.66
E_1·MGT	− 3066.00	149.39	− 20.52
E_2·MGT	1836.50	131.22	13.99
CONSTANT	11204.00	79.10	141.64
$n = 46$	$R^2 = .9988$	$s = 173.88$	

Table 4.4. Regression analysis of salary data: Expanded model, obs. 33 deleted

Variable	Coefficient	SE	t
EXPRNC	498.40	2.16	231.06
E_1	−1741.00	40.78	−42.69
E_2	−356.90	37.77	−9.45
MGT	7040.50	39.72	177.25
$E_1 \cdot$ MGT	−3051.00	57.82	−52.77
$E_2 \cdot$ MGT	1997.50	51.91	38.48
CONSTANT	11200.00	30.61	365.89
$n = 45$	$R^2 = .9998$	$s = 67.28$	

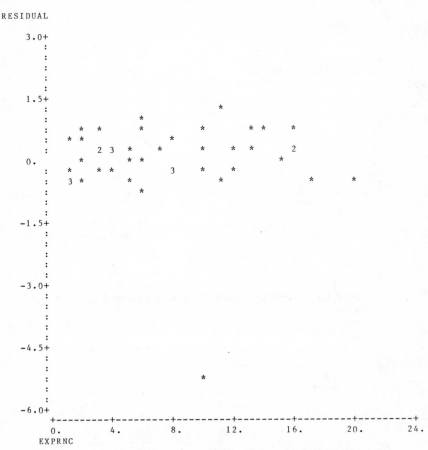

Fig. 4.4. Standardized residuals versus years of experience: Expanded model.

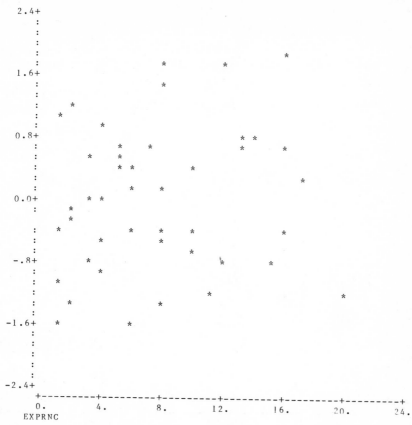

Fig. 4.5. Standardized residuals versus years of experience: Expanded model, obs. 33 deleted.

each of these groups has residuals that appear to be symmetrically distributed about zero. Therefore the introduction of the interaction terms has produced an accurate representation of salary variations. The relationship between income, experience, education, and managerial status appears to be adequately described by the model given in Equation (4.3). The model appears to have acceptable specification.

With the standard error of the residuals estimated to be $67.28, we can believe that we have uncovered the actual and very carefully administered salary formula. Using 95% confidence intervals, each year of experience is estimated to be worth between $494.08 and $502.72. These increments of approximately $500 are added to a starting salary that is specified for each of the six education–management groups. Since the final regression model

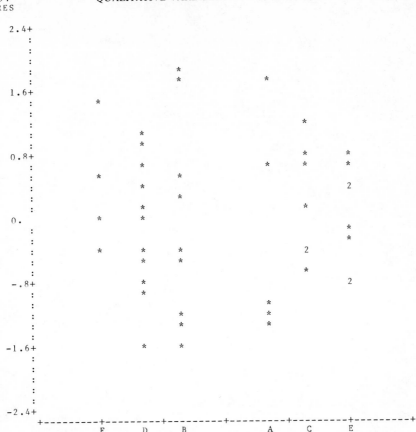

Fig. 4.6. Standardized residuals versus education–management categorical variable: Expanded model, obs. 33 deleted.

is not additive, it is rather difficult to directly interpret the coefficients of the dummy variables. In order to see how the qualitative variables affect salary differentials, we use the coefficients to form estimates of the base salary for each of the six categories. These results are in Table 4.5 along with standard errors and confidence intervals. The standard errors are computed using Equation (3.8) of Chapter 3.

Using a regression model with dummy variables and interaction terms, it has been possible to explain almost all the variation in salaries of computer professionals selected for this survey. The level of accuracy with which the model explains the data is very rare! We can only conjecture that the

Table 4.5.   Estimates of base salary using nonadditive model

$$Y = \beta_0 + \beta_1 EXPRNC + \gamma_1 E_1 + \gamma_2 E_2 + \delta_1 MGT$$
$$+ \alpha_1 E_1 \cdot MGT + \alpha_2 E_2 \cdot MGT$$

Category		Coefficients	Estimate of base salary[a]	SE[a]	95% Confidence interval
EDUC	MGT				
1	1	$\beta_0 + \gamma_1 + \delta + \alpha_1$	13448	32	13385–13511
1	0	$\beta_0 + \gamma_1$	9459	31	9398– 9520
2	1	$\beta_0 + \gamma_2 + \delta + \alpha_2$	19880	33	19815–19945
2	0	$\beta_0 + \gamma_2$	10843	26	10792–10894
3	1	$\beta_0 + \delta$	18240	29	18183–18297
3	0	β_0	11200	31	11139–11261

[a]Recorded to the nearest dollar.

methods of salary administration in this company are precisely defined and strictly applied.

In retrospect, we see that an equivalent model may be obtained with a different set of dummy variables and regression parameters. One could define five variables, each taking on the values of 1 or 0 corresponding to five of the six education–management categories. The numerical estimates of base salary and the standard errors of Table 4.5 would be the same. The advantage to proceeding as we have is that it allows us to separate the effects of the three sets of explanatory variables, (i) education, (ii) management, and (iii) education–management interaction. Recall that interaction terms were included only after we found that an additive model did not satisfactorily explain salary variations. In general, we start with simple models and proceed sequentially to more complex models if necessary. We shall always hope to retain the simplest model that has an acceptable residual structure.

4.3.   SYSTEMS OF REGRESSION EQUATIONS: COMPARING TWO GROUPS

A collection of data may consist of two or more distinct subsets, each of which may require a separate regression equation. Serious bias may be incurred if one regression relationship is used to represent the pooled data set. An analysis of this problem can be accomplished using indicator variables. An analysis of separate regression equations for subsets of the data may be applied to cross section or time series data. The example discussed below treats cross section data. Applications to time series data are discussed in the final section of this chapter.

Pre-employment Tests and Job Performance: Two Groups

The first example is an analysis of data that is associated with a currently important problem concerning equal opportunity in employment. Many large corporations and government agencies administer a pre-employment test in an attempt to screen job applicants. The test is supposed to measure an applicant's aptitude for the job and the results are used as part of the information for making a hiring decision. The federal government has ruled* that these tests (i) must measure abilities that are directly related to the job under consideration and (ii) must not discriminate on the basis of race or national origin. Operational definitions of requirements (i) and (ii) are rather elusive. We shall not try to resolve these operational problems. We shall take one approach involving race represented as two groups, white and minority. The hypothesis that there are separate regressions relating test scores to job performance for the two groups will be examined. The implications of this hypothesis for discrimination in hiring are discussed.

Let Y represent job performance and let X be the score on the pre-employment test. We want to compare

$$\text{Model 1:} \quad Y_{ij} = \beta_0 + \beta_1 X_{ij} + u_{ij} \qquad \text{(Pooled)}$$
$$i = 1,2; \qquad j = 1,2,\ldots,n_i$$
$$\text{Model 2:} \quad Y_{1j} = \beta_{10} + \beta_{11} X_{1j} + u_{1j} \qquad \text{(Minority)}$$
$$Y_{2j} = \beta_{20} + \beta_{21} X_{2j} + u_{2j} \qquad \text{(White)}$$

Figure 4.7 depicts the two models. In Model 1, race distinction is ignored, the data is pooled and there is one regression line. In Model 2 there is a separate regression relationship for the two subgroups, each with distinct regression coefficients. We shall assume that the variances of the residual terms are the same in each subgroup.

Before analyzing the data, let us briefly consider the type of errors that could be present in interpreting and applying the results. If Y_0 as seen on the graph, has been set as the minimum required level of performance, then using Model 1, an acceptable score on the test is one that exceeds X_p. However, if Model 2 is in fact correct, the appropriate test score for whites is X_w and for minorities it is X_m. Using X_p in place of X_m and X_w represents a relaxation of the pretest requirement for whites and a tightening of that requirement for minorities. Since inequities can result in the selection procedure if the wrong model is used to set cutoff values, it is necessary to

*Tower amendment to Title VII, Civil Rights Act of 1964.

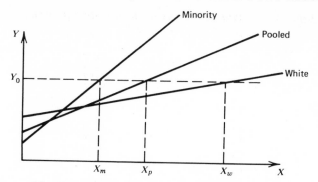

Fig. 4.7.   Requirements for employment on pre-test.

examine the data carefully. It must be determined whether there are two distinct relationships or whether the relationship is the same for both groups and a single equation estimated from the pooled data is adequate. Note that whether Model 1 or Model 2 is chosen, the values X_m, X_w, X_p are estimates subject to sampling error and should only be used in conjunction with appropriate confidence intervals. (Construction of confidence intervals is discussed in the following paragraphs.)

Data was collected for this analysis using a special employment program.* Twenty applicants were hired on a trial basis for 6 weeks. One week was spent in a training class. The remaining 5 weeks were spent on the job. The participants were selected from a pool of applicants by a method that was not related to the pre-employment test scores. A test was given at the end of the training period and a work performance evaluation was developed at the end of the 6-week period. These two scores were combined to form an index of job performance. (Those employees with unsatisfactory performance at the end of 6 weeks were dropped.) The data appears in Table 4.6.

Formally we want to test the null hypothesis, H: $\beta_{11} = \beta_{21}$, $\beta_{10} = \beta_{20}$ against the alternative that there are substantial differences in the β's. The test can be performed using indicator variables. Let Z_{ij} be defined to take the value 1 if $i = 1$ and to take the value zero if $i = 2$. That is, Z is a new variable that has the value 1 for each observation on a minority applicant and the value zero for each observation on a white applicant. Then we consider the two models,

$$\text{Model 1:} \qquad Y_{ij} = \beta_0 + \beta_1 X_{ij} + u_{ij}$$

*The data is not based on an actual experiment. It was constructed by the authors.

Table 4.6.   Data on pre-employment testing program

	RACE	JPERF	TEST
001	1	1.83	0.28
002	1	4.59	0.97
003	1	2.97	1.25
004	1	8.14	2.46
005	1	8.00	2.51
006	1	3.30	1.17
007	1	7.53	1.78
008	1	2.03	1.21
009	1	5.00	1.63
010	1	8.04	1.98
011	0	3.25	2.36
012	0	5.30	2.11
013	0	1.39	0.45
014	0	4.69	1.76
015	0	6.56	2.09
016	0	3.00	1.50
017	0	5.85	1.25
018	0	1.90	0.72
019	0	3.85	0.42
020	0	2.95	1.53

Source: Created by the authors.

and

Model 3: $Y_{ij} = \beta_0 + \beta_1 X_{ij} + \gamma Z_{ij} + \delta (Z_{ij} \cdot X_{ij}) + u_{ij}.$

Note that Model 3 is equivalent to Model 2 if we observe that

$$\beta_0 = \beta_{20}, \quad \beta_{10} - \beta_{20} = \gamma,$$
$$\beta_1 = \beta_{21}, \quad \beta_{11} - \beta_{21} = \delta.$$

Our null hypothesis H, now becomes $H: \gamma = \delta = 0$. The hypothesis is tested by constructing an F statistic for the comparison of two models as described in Chapter 3. In this case,

$$F = \frac{(\text{SSE(RM)} - \text{SSE(FM)})/2}{\text{SSE(FM)}/16},$$

where the full model (FM) is Model 3 and the restricted model (RM) is Model 1, and F has 2 and 16 degrees of freedom. Proceeding with the analysis of the data, the regression results for Model 1 and Model 3 are given in Tables 4.7 and 4.8. The plots of residuals against the explanatory variables (Figures 4.8, 4.9) look acceptable in both cases. The one residual to the lower right in Model 1 may require further investigation.

Table 4.7. Regression results, pre-employment testing data: Model 1

Variable	Coefficient	SE	t
TEST (X)	2.36	.54	4.37
CONSTANT	1.03	.87	1.18
$n = 20$	$R^2 = .517$	$s = 1.59$	

Table 4.8. Regression results, pre-employment testing data: Model 3

Variable	Coefficient	SE	t
RACE (Z)	-1.91	1.54	-1.24
RACE·TEST ($X·Z$)	1.99	.95	2.09
TEST (X)	1.32	.67	1.95
CONSTANT	2.01	1.05	1.91
$n = 20$	$R^2 = .664$	$s = 1.41$	

To evaluate the formal hypothesis we compute the F ratio specified previously which is equal to

$$F = \frac{(45.51 - 31.81)/2}{31.81/16} = 3.4$$

and is significant at a level slightly above 5%. Therefore on the basis of this test we would conclude that the relationship is probably different for the two groups. Specifically, for whites we have

$$Y_2 = 2.01 + 1.32X_2$$

and for minority races

$$Y_1 = .10 + 3.31X_1.$$

The results are very similar to those that were described earlier in Figure 4.6 when the problem of bias was discussed. The straight line representing the relationship for minorities has a larger slope and a smaller intercept than the line for whites. If a pooled model were used, the types of biases discussed in relation to Figure 4.7 would occur.

Although the formal procedure using dummy variables has led to the plausible conclusion that the relationships are different for the two groups, the data for the individual groups has not been looked at carefully. Recall

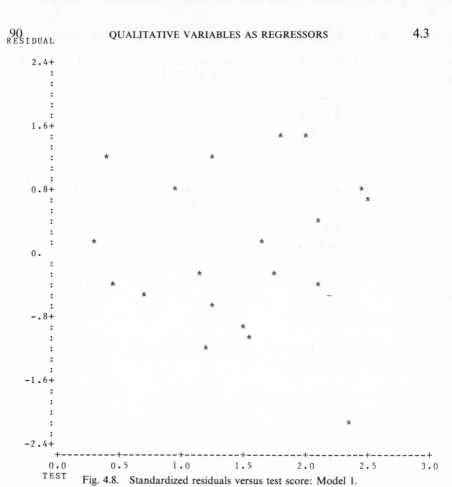

Fig. 4.8.   Standardized residuals versus test score: Model 1.

that it was assumed that the variances were identical in the two groups. This assumption was required so that the only distinguishing characteristic between the two samples was the pair of regression coefficients. In Figure 4.10 a plot of residuals versus the indicator variable is presented. There does not appear to be a difference between the two sets of residuals. We shall now look more closely at each group. The regression coefficients for each sample taken separately are in Table 4.9. The residuals are in Figures 4.11 and 4.12. The regression coefficients are, of course, the values obtained from Model 3. The standard errors of the residuals are 1.29 and 1.51 for the minority and white samples, respectively. The residual plots against the test score are acceptable in both cases. An interesting observation that was not available in the earlier analysis is that the pre-employment test

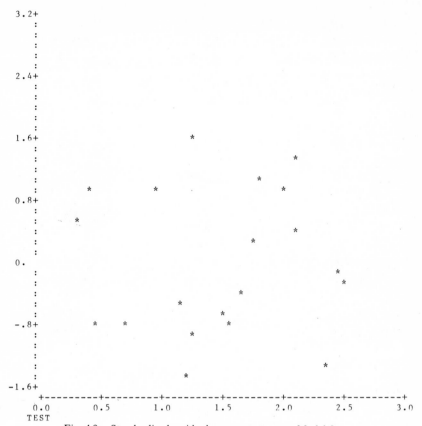

Fig. 4.9.   Standardized residuals versus test score: Model 3.

explains a major portion of the variation in the minority sample, but the test is only marginally useful in the white sample.

Our previous conclusion is still valid. The two regression equations are different. Not only are the regression coefficients different, but the standard deviations of the residuals also show slight differences. Of more importance, the values of R^2 are greatly different. For the white sample, $R^2 = .29$ is so small ($t = 1.82$; 2.306 is required for significance) that the pre-employment test score is not deemed an adequate predictor of job success. This finding has bearing on our original objective since it should be a prerequisite for comparing regressions in two samples that the relationships be valid in each of the samples when taken alone. Concerning the validity of the pre-employment test, we conclude that if applied as the

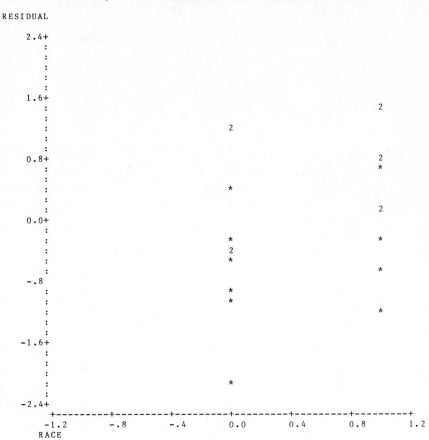

Fig. 4.10.   Standardized residuals versus race: Model 1.

law prescribes, with indifference to race, it will give biased results for both racial groups. Moreover, based on these findings we may be justified in saying that the test is of no value for screening white applicants.

We close the discussion with a note about determining the appropriate cutoff test score if the test were used. Consider the results for the minority sample. If Y_m is designated as the minimum acceptable job performance value to be considered successful, then from the regression equation (also see Figure 4.7)

$$X_m = \frac{Y_m - b_0}{b_1},$$

where b_0 and b_1 are the estimated regression coefficients. X_m is an estimate

Table 4.9.   *Separate regression results*

Sample	b_0	b_1	t_1	R^2	SD[a] of residuals
White	2.01	1.32	1.82	.29	1.51
Minority	.10	3.31	5.30	.78	1.29

[a]Standard deviation.

of the minimum acceptable test score required to attain Y_m. Since X_m is defined in terms of quantities with sampling variation, X_m is also subject to sampling variation. The variation is most easily summarized by constructing a confidence interval for X_m. An approximate 95% level confidence

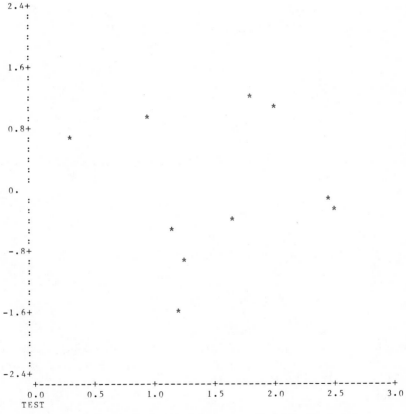

Fig. 4.11. Standardized residuals versus test: Model 1, minority only.

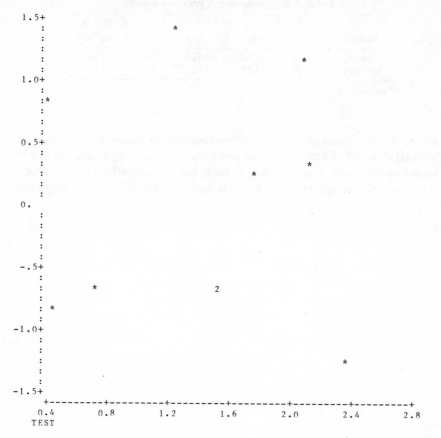

Fig. 4.12.   Standardized residuals versus test: Model 1, white only.

interval takes the form (Scheffé, 1959, page 52)

$$X_m - \frac{ts/n}{b_1}, \qquad X_m + \frac{ts/n}{b_1},$$

where t is the appropriate percentile point of the "t" distribution, and s is the standard deviation of residuals. If Y_m is set at 4, then $X_m = (4 - .10)/3.31 = 1.18$ and a 95% confidence interval for the test cutoff score is 1.09–1.27.

4.4. DUMMY VARIABLES: OTHER APPLICATIONS

Applications of dummy variables such as those described in the last two examples can be extended to cover a variety of problems. Both the number of quantitative explanatory variables and the number of distinct groups represented in the data by dummy variables may be increased. (See Kerlinger and Pedhazur, 1973; Kmenta, 1971; and Searle, 1971 for a variety of applications.) Dummy variables can also be utilized with time series data. (Note that both examples discussed above are based on cross section data.) In addition, there are some models of growth processes where an indicator variable is used as the dependent variable.

In the next two sections we discuss the use of dummy variables with time series data. In particular, notions of seasonality and stability of parameters over time will be discussed. These problems are formulated and the data is provided. The analyses are left to the reader.

4.5. SEASONALITY

In Chapter 6, sales data for a firm that manufactures skis and related equipment is analyzed. The model is an equation that relates sales to personal disposable income (PDI),* each variable measured quarterly. The basic model is

$$S_t = \beta_0 + \beta_1 \cdot \text{PDI}_t + u_t,$$

where S_t is sales in millions in the tth period. The data appears in Table 6.6 of Chapter 6. Our approach here is to assume the existence of a seasonal effect on sales that is determined on a quarterly basis. To measure this effect we may define indicator variables to characterize the seasonality. In particular we begin with

$$Z_{1t} = \begin{matrix} 1 \\ 0 \end{matrix} \quad \begin{matrix} \text{if the } t\text{th period is a first quarter} \\ \text{otherwise,} \end{matrix}$$

$$Z_{2t} = \begin{matrix} 1 \\ 0 \end{matrix} \quad \begin{matrix} \text{if the } t\text{th period is a second quarter} \\ \text{otherwise,} \end{matrix}$$

$$Z_{3t} = \begin{matrix} 1 \\ 0 \end{matrix} \quad \begin{matrix} \text{if the } t\text{th period is a third quarter} \\ \text{otherwise.} \end{matrix}$$

The analysis and interpretation is left to the reader. The authors have

---

*Aggregate measure of purchasing potential.

analyzed this data and found that there are actually only two seasons. (See the discussion of this sales data in Chapter 6 for an analysis using only one indicator variable, two seasons.) Also see Kmenta (1971) for further discussion on using indicator variables for analyzing seasonality.

4.6. STABILITY OF REGRESSION PARAMETERS OVER TIME

Dummy variables may also be used to analyze the stability of regression coefficients over time or to test for structural change. We consider an extension of the system of regressions problem when data is available on a cross section of observations and over time. Our objective is to analyze the constancy of the relationships over time. The methods described here are suitable for intertemporal and interspatial comparisons. In order to outline the method we refer to a data set which is used in Chapter 5 to demonstrate methods of dealing with heteroscedasticity in multiple regression. The proposed model specifies that per capita expenditure on public education in a state, Y, can be explained using three variables: X_1–per capita personal income, X_2–proportion of population under 18 years of age, and X_3–proportion of population residing in urban areas. In Chapter 5 the effects of regional characteristics on the regression relationship are analyzed. In this section we focus on the stability of the expenditure relationship with respect to time.

Data have been developed on the four variables previously described for each state in 1965, 1970, and 1975. Assuming that the relationship can be identically specified in each of the 3 years,* the analysis of stability can be carried out by evaluating the variation in the estimated regression coefficients over time. Working with the pooled data set of 150 observations (50 states each in 3 years) we define indicator variables

$$T_{1i} = \begin{matrix} 1 \\ 0 \end{matrix} \quad \begin{matrix} \text{if } i\text{th observation was from 1965} \\ \text{otherwise,} \end{matrix}$$

and

$$T_{2i} = \begin{matrix} 1 \\ 0 \end{matrix} \quad \begin{matrix} \text{if the } i\text{th observation was from 1970} \\ \text{otherwise.} \end{matrix}$$

Using Y to represent per capita expenditure on schools, the model takes

---

*Specification as used here means that the same variables appear in each equation. Any transformations that are used apply to each equation. The assumption concerning identical specification should be empirically validated.

Table 4.10.   Education expenditures

CASE	STATE	URB60	SE60	PI58	Y59	GEOG
1	ME	399	61	1704	388	1
2	NH	598	68	1885	372	1
3	VT	370	72	1745	397	1
4	MA	868	72	2394	358	1
5	RI	899	62	1966	357	1
6	CT	690	91	2817	362	1
7	NY	728	104	2685	341	1
8	NJ	826	99	2521	353	1
9	PA	656	70	2127	352	1
10	OH	674	82	2184	387	2
11	IN	568	84	1990	392	2
12	IL	759	84	2435	366	2
13	MI	650	104	2099	403	2
14	WI	621	84	1936	393	2
15	MN	610	103	1916	402	2
16	IA	522	86	1863	385	2
17	MO	613	69	2037	364	2
18	ND	351	94	1697	429	2
19	SD	390	79	1644	411	2
20	NB	520	80	1894	379	2
21	KS	564	98	2001	380	2
22	DE	326	124	2760	388	3
23	MD	562	92	2221	393	3
24	VA	487	67	1674	402	3
25	WV	358	66	1509	405	3
26	NC	362	65	1384	423	3
27	SC	343	57	1218	453	3
28	GA	498	60	1487	420	3
29	FL	628	74	1876	334	3
30	KY	377	49	1397	594	3
31	TN	457	60	1439	346	3
32	AL	517	59	1359	637	3
33	MS	362	68	1053	448	3
34	AR	416	56	1225	403	3
35	LA	562	72	1576	433	3
36	OK	610	80	1740	378	3
37	TX	727	79	1814	409	3
38	MT	463	95	1920	412	4
39	ID	414	79	1701	418	4
40	WY	568	142	2088	415	4
41	CO	621	108	2047	399	4
42	NM	618	94	1838	458	4
43	AZ	699	107	1932	425	4
44	UT	665	109	1753	494	4
45	NV	663	114	2569	372	4
46	WA	584	112	2160	386	4
47	OR	534	105	2006	382	4
48	CA	717	129	2557	373	4
49	AK	379	107	1900	434	4
50	HI	693	77	1852	431	4

CASE	STATE	URB70	SE70	PI68	Y69	GEOG
1	ME	508	189	2828	351	1
2	NH	564	169	3259	346	1
3	VT	322	230	3072	348	1
4	MA	846	168	3835	335	1
5	RI	871	180	3549	327	1
6	CT	774	193	4256	341	1
7	NY	856	261	4151	326	1
8	NJ	889	214	3954	333	1
9	PA	715	201	3419	326	1
10	OH	753	172	3509	354	2
11	IN	649	194	3412	359	2
12	IL	830	189	3981	349	2
13	MI	738	233	3675	369	2
14	WI	659	209	3363	361	2
15	MN	664	262	3341	365	2
16	IA	572	234	3265	344	2
17	MO	701	177	3257	336	2
18	ND	443	177	2730	369	2
19	SD	446	187	2876	369	2
20	NB	615	148	3239	350	2
21	KS	661	196	3303	340	2
22	DE	722	248	3795	376	3
23	MD	766	247	3742	364	3
24	VA	631	180	3068	353	3
25	WV	390	149	2470	329	3
26	NC	450	155	2664	354	3
27	SC	476	149	2380	377	3
28	GA	603	156	2781	371	3
29	FL	805	191	3191	336	3
30	KY	523	140	2645	349	3
31	TN	588	137	2579	343	3
32	AL	584	112	2337	362	3
33	MS	445	130	2081	385	3
34	AR	500	134	2322	352	3
35	LA	661	162	2634	390	3
36	OK	680	135	2880	330	3
37	TX	797	155	3029	369	3
38	MT	534	238	2942	369	4
39	ID	541	170	2668	368	4
40	WY	605	238	3190	366	4
41	CO	785	192	3340	358	4
42	NM	698	227	2651	421	4
43	AZ	796	207	3027	387	4
44	UT	804	201	2790	412	4
45	NV	809	225	3957	385	4
46	WA	726	215	3688	342	4
47	OR	671	233	3317	333	4
48	CA	909	273	3968	348	4
49	AK	484	372	4146	440	4
50	HI	831	212	3513	383	4

CASE	STATE	URB70	SE75	PI73	Y74	GEOG
1	ME	508	235	3944	325	1
2	NH	564	231	4578	323	1
3	VT	322	270	4011	328	1
4	MA	846	261	5233	305	1
5	RI	871	300	4780	303	1
6	CT	774	317	5889	307	1
7	NY	856	387	5663	301	1
8	NJ	889	285	5759	310	1
9	PA	715	300	4894	300	1
10	OH	753	221	5012	324	2
11	IN	649	264	4908	329	2
12	IL	830	308	5753	320	2
13	MI	738	379	5439	337	2
14	WI	659	342	4634	328	2
15	MN	664	378	4921	330	2
16	IA	572	232	4869	318	2
17	MO	701	231	4672	309	2
18	ND	443	246	4782	333	2
19	SD	446	230	4296	330	2
20	NB	615	268	4827	318	2
21	KS	661	337	5057	304	2
22	DE	722	344	5540	328	3
23	MD	766	330	5331	323	3
24	VA	631	261	4715	317	3
25	WV	390	214	3828	310	3
26	NC	450	245	4120	321	3
27	SC	476	233	3817	342	3
28	GA	603	250	4243	339	3
29	FL	805	243	4647	287	3
30	KY	523	216	3967	325	3
31	TN	588	212	3946	315	3
32	AL	584	208	3724	332	3
33	MS	445	215	3448	358	3
34	AR	500	221	3680	320	3
35	LA	661	244	3825	355	3
36	OK	680	234	4189	306	3
37	TX	797	269	4336	335	3
38	MT	534	302	4418	335	4
39	ID	541	268	4323	344	4
40	WY	605	323	4813	331	4
41	CO	785	304	5046	324	4
42	NM	698	317	3764	366	4
43	AZ	796	332	4504	340	4
44	UT	804	315	4005	378	4
45	NV	809	291	5560	330	4
46	WA	726	312	4989	313	4
47	OR	671	316	4697	305	4
48	CA	909	332	5438	307	4
49	AK	484	546	5613	386	4
50	HI	831	311	5309	333	4

PIi   Per capita income in year i

Yi   Number of residents per thousand under 18 years of age in year i

URBi   Number of residents per thousand living in urban areas in year i

SEi   Per capita expenditure on education for year i

the form

$$Y = \beta_0 + \beta_1 X_1 + \beta_2 X_2 + \beta_3 X_3 + \gamma_1 T_1 + \gamma_2 T_2$$
$$+ \delta_1 T_1 \cdot X_1 + \delta_2 T_1 \cdot X_2 + \delta_3 T_1 \cdot X_3$$
$$+ \Phi_1 T_2 \cdot X_1 + \Phi_2 T_2 \cdot X_2 + \Phi_3 T_2 \cdot X_3 + u.$$

From the definitions of T_1 and T_2, the previous model is equivalent to

For 1965: $Y = (\beta_0 + \gamma_1) + (\beta_1 + \delta_1) X_1 + (\beta_2 + \delta_2) X_2 + (\beta_3 + \delta_3) X_3 + u,$

For 1970: $Y = (\beta_0 + \gamma_2) + (\beta_1 + \Phi_1) X_1 + (\beta_2 + \Phi_2) X_2 + (\beta_3 + \Phi_3) X_3 + u,$

For 1975: $Y = \beta_0 + \beta_1 X_1 + \beta_2 X_2 + \beta_3 X_3 + u.$

As noted earlier, this method of analysis necessarily implies that the variability about the regression function is assumed to be equal for all 3 years. One formal hypothesis of interest is

$$H: \quad \gamma_1 = \gamma_2 = \delta_1 = \delta_2 = \delta_3 = \Phi_1 = \Phi_2 = \Phi_3 = 0,$$

which implies that the regression system has remained unchanged throughout the period of investigation (1965–1975).

The data for this example appears in Table 4.10. The reader is invited to perform the analysis described above as an exercise.

REFERENCES

Kerlinger, F. N. and E. J. Pedhazur, *Multiple Regression in Behavioral Research*, Holt, Rinehart and Winston, New York, 1973.

Kmenta, J., *Elements of Econometrics*, Macmillan, New York, 1971.

Scheffé, H., *The Analysis of Variance*, Wiley, New York, 1959.

Searle, S. R., *Linear Models*, Wiley, New York, 1971.

CHAPTER 5

Weighted Least Squares

5.1. INTRODUCTION

In the preceding chapters, 1 through 4, it has been assumed that the underlying correct regression model is of the form

$$Y_i = \beta_0 + \beta_1 X_{1i} + \cdots + \beta_p X_{pi} + u_i, \qquad (5.1)$$

where u_i's are random disturbances that are independent and identically distributed (i.i.d.). Various residual plots have been used to check these assumptions. If the residuals are not consistent with the assumptions, it is suggested that either the equation form is inadequate, some additional variables are required, or some of the data observations are outliers.

There has been one exception to this line of analysis. In the example based on the Supervisor data of Chapter 2, it was argued that the underlying model did not have residuals that were i.i.d. In particular, the residuals did not have constant variance. This situation (nonconstant residual variance) is often referred to as heteroscedasticity. The presence of unequal variances violates one of the basic ordinary least squares (OLS) assumptions. If OLS is applied, ignoring heteroscedasticity, the estimated coefficients are still unbiased, but are no longer best in the sense of precision (variance). For the Supervisor data, a transformation was imposed to correct the situation so that better estimates of the original model parameters could be obtained (better than OLS).

In this chapter and the one that follows, we investigate some regression situations where the underlying process implies that the regression residuals are not i.i.d. In the present chapter, heteroscedasticity is discussed. The problem is resolved by applying variations of weighted least squares (WLS). In the next chapter regression models with residuals that are not independent are treated. The approach in both situations is to use a combination of prior knowledge, intuition, and evidence found in the OLS residuals to detect the problem. The solution is usually prescribed as a two-stage procedure. In stage 1, the OLS residuals are used to estimate the parameters of the residual structure. In the second stage, these estimates are used to define a transformation or procedure that corrects for the lack of i.i.d. residuals and to produce estimates of the regression coefficients that usually have more precision than the OLS estimates.

5.2. HETEROSCEDASTIC MODELS

Three different heteroscedastic situations will be distinguished. The first two situations are fairly simple. In these two cases, once the necessity for WLS has been recognized, estimation can be accomplished in one step. The third situation is more complex and requires a two-stage estimation procedure. An example of the first heteroscedastic situation is found in Chapter 2 and will be reviewed here. The second situation is formulated, but no data is analyzed. The third heteroscedastic situation is demonstrated with two examples.

5.3. SUPERVISOR DATA

The first heteroscedastic situation has been treated in Chapter 2. There, data on X, the number of workers in an industrial establishment, and Y, the number of supervisors in the establishment were presented for 27 establishments. The regression model was

$$Y_i = \beta_0 + \beta_1 X_i + u_i. \tag{5.2}$$

It was argued that the variance of u_i depends on the size of the establishment as measured by X; that is, $\sigma_{u_i}^2 = k^2 X_i^2$ where k is a positive constant. (See Chapter 2 for details.) Empirical evidence for this type of heteroscedasticity is obtained by plotting the OLS residuals against X. A plot with the characteristics of Figure 5.1 typifies the situation. If corrective action is not taken and OLS is applied to the raw data, the resulting estimated coefficients will lack precision in a theoretical sense. In addition, for the type of heteroscedasticity present in this data, the estimated standard errors of the regression coefficients are often understated giving a false sense of precision. The problem is resolved by using a version of weighted least squares as described in Chapter 2.

This approach to heteroscedasticity may also be considered in multiple regression models. In Equation (5.1) the variance of the residuals may be affected by only one of the explanatory variables. (The case where the variance is a function of more than one explanatory variable is discussed

$$\sigma_u = kX$$

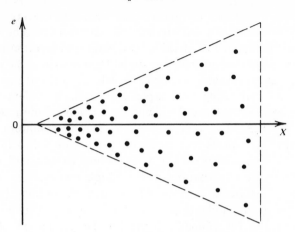

Fig. 5.1. Heteroscedastic residuals.

later.) Empirical evidence is available from the plot of OLS residuals versus the suspected variable and correction is accomplished by extending the method applied in Chapter 2. The resulting estimates are obtained by a transformation of the data. For example, if the original model is given as Equation (5.1) and it is found that $\sigma_{u_i} = kX_{4i}$, then the estimates are produced by regressing Y_i/X_{4i} against $1/X_{4i}, X_{1i}/X_{4i}, \ldots, X_{3i}/X_{4i}, X_{5i}/X_{4i}, \ldots, X_{pi}/X_{4i}$. The resulting coefficient of $1/X_{4i}$ is b_0, an estimate of β_0, the coefficient of X_{1i}/X_{4i} is an estimate of β_1, and so on, and the intercept from the regression is an estimate of β_4. Refer to Chapter 2 for a detailed discussion of this method as applied in simple regression.

5.4. COLLEGE EXPENSE DATA

A second heteroscedastic situation arises frequently with large-scale survey data where measurements on individual sampling units are averaged over a well-defined cluster of units in order to obtain increased stability. Only the average and number of sampling units are reported as data. For example, consider a survey of undergraduate college students (or their parents) that is intended to assess total annual college-related expenses. Assume that the survey is also intended to collect information that will make it possible to relate expenses to characteristics of the institution attended. Regression analysis may be used with a model such as

$$Y_i = \beta_0 + \beta_1 X_{1i} + \beta_2 X_{2i} + \cdots + \beta_6 X_{6i} + u_i. \tag{5.3}$$

The variables are defined in Table 5.1. The data may be collected by selecting a set of schools at random and then interviewing a prescribed number of randomly selected students at each school. The explanatory variables are characteristics of the school with the exception of X_6, which can be taken as an average over the student population. (The logic behind choosing these explanatory variables is left to the imagination of the reader.) Rather than using total expense Y for each student interviewed, the average expense for these students at each institution serves as the dependent variable. The precision of average expenditure is directly proportional to the square root of the sample size on which the average is based. That is, the variance of $\bar{Y}$ is σ^2/n and its standard deviation is $\sigma/\sqrt{n}$. If there are k institutions in the sample and $n_1, n_2, \ldots, n_k$ represent the number of students interviewed at each institution, the standard deviation of u_i in the model (Equation (5.1)) is $\sigma_{u_i} = \sigma/\sqrt{n_i}$ where σ is the standard deviation for annual expense for the population of individual students. Estimation of the regression coefficients is carried out using WLS with weights $w_i = 1/\sigma_{u_i}^2$ as in Chapter 2. Since $\sigma_{u_i}^2 = \sigma^2/n_i$, the regression coefficients are obtained by minimizing the weighted sum of squared residuals,

$$S = \sum_{i=1}^{k} n_i \left(Y_i - \beta_0 - \sum_{j=1}^{6} \beta_j X_{ji} \right)^2. \qquad (5.4)$$

Note that the procedure implicitly recognizes that observations from institutions where a large number of students were interviewed are more reliable and should have more weight in determining the regression coefficients than observations from institutions where only a few students were interviewed. The differential precision associated with different observation may be taken as a justification for the weighting scheme.

The estimated coefficients and summary statistics may be computed

Table 5.1. *Variables in cost of education survey*

Name	Description
Y	Total annual expense (above tuition)
X_1	Size of city or town where school is located
X_2	Distance to nearest urban center
X_3	Type of school—public, private
X_4	Size of student body
X_5	Proportion of entering freshman that graduate
X_6	Distance from home

using a special WLS computer program or by transforming the data and using OLS as in the example in Chapter 2. If both sides of Equation (5.1) are multiplied by $n_i^{1/2}$, the new model will have residuals, $\epsilon_i = u_i \cdot n_i^{1/2}$ and $\sigma_{\epsilon_i} = \sigma$, a constant. That is, the regression model stated in the new variables is

$$Y_i n_i^{1/2} = \beta_0 n_i^{1/2} + \beta_1 X_{1i} n_i^{1/2} + \cdots + \beta_6 X_{6i} n_i^{1/2} + \epsilon_i. \qquad (5.5)$$

The residuals in Equation (5.5) satisfy the necessary assumption of constant variance. Regression of $Y_i \cdot n_i^{1/2}$ against the seven new variables consisting of $n_i^{1/2}$, and the six transformed explanatory variables, $X_{ji} n_i^{1/2}$ using OLS will produce the desired estimates of the regression coefficients and their standard errors. Note that the regression with the transformed variables must be carried out with the constant term constrained to be zero. That is, β_0, the intercept of the original model is now the coefficient of $n_i^{1/2}$. Equation (5.5) has no intercept. More details on this point are given with the numerical example in section 5.6.

5.5. TWO-STAGE ESTIMATION

In the two preceding problems heteroscedasticity was expected at the outset. In the first problem the nature of the process under investigation suggests residual variances that increase with the size of the explanatory variable. In the second case, the method of data collection indicates heteroscedasticity. In both cases, homogeneity of variance is accomplished by a transformation. The transformation is constructed directly from information in the raw data. In the problem described in this section, there is also some prior indication that the variances are not equal. But here the exact structure of heteroscedasticity is determined empirically. As a result, estimation of the regression parameters requires two stages.

It is not a simple matter to detect heteroscedasticity in a general multiple regression situation. If present it is often discovered as a result of some good intuition on the part of the analyst on how observations may be grouped or clustered. For multiple regression models, the plot of residuals against $\hat{Y}_i$, the fitted values of the response variable, can serve as a first step. If the magnitude of the residuals appears to vary systematically with $\hat{Y}_i$, heteroscedasticity is suggested. The plot does not necessarily clearly identify the source of the problem. (See the following example.)

One direct method for investigating the presence of nonconstant variance is available when there are replicated measurements on the response variable corresponding to a set of fixed values of the explanatory variables. For example, in the case of one explanatory variable, we may have measurements $Y_{11}, Y_{12}, \ldots, Y_{1n_1}$ at X_1; $Y_{21}, Y_{22}, \ldots, Y_{2n_2}$ at X_2; and so on up to $Y_{k1}, Y_{k2}, \ldots, Y_{kn_k}$ at X_k. Taking $k = 4$ for illustrative purposes, a plot of the data appears as Figure 5.2. With this wealth of data, it is not necessary to make restrictive assumptions regarding the nature of heteroscedasticity such as $\sigma_{u_i} = kX_i$. It is clear from the graph that the nonconstancy of variance does not follow a simple systematic pattern. The variability first decreases as X increases, up to X_3 then jumps again at X_4. The regression model could be stated as

$$Y_{ij} = \beta_0 + \beta_1 X_{ij} + u_{ij}, \quad i = 1, 2, 3, 4;$$
$$X_{ij} \equiv X_i \qquad \sigma_{u_{ij}} = \sigma_i, \quad j = 1, 2, \ldots, n_i. \tag{5.6}$$

Each observed residual, e_{ij}, is made up of two parts, the difference between Y_{ij} and $\bar{Y}_i$ and the difference between $\bar{Y}_i$ and the point on the regression line, $\hat{Y}_i$. That is, $e_{ij} = (Y_{ij} - \bar{Y}_i) + (\bar{Y}_i - \hat{Y}_{ij})$. The first part is referred to as pure error. The second part measures lack of fit. An assessment of heteroscedasticity is based on pure error.* Weights for WLS may be estimated as $w_i = \dfrac{1}{s_i^2}$, where $s_i^2 = \sum (Y_{i_j} - \bar{Y}_i)^2 / (n_i - 1)$.

The presence of replications on the response variable for a given value of X is rather uncommon when data are collected in a nonexperimental setting. (When the data are collected in a controlled laboratory setting, the researcher can choose to replicate any observations.) When there is only one explanatory variable, it is possible that some replications will occur. If

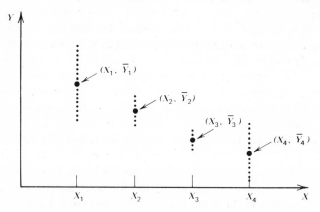

Fig. 5.2.   Non constant variance with replicated observations.

---

*The notion of pure error can also be used to obtain a test for lack of fit. See Draper and Smith (1966).

there are many explanatory variables, it is virtually impossible to imagine coming upon two observations with identical values on all explanatory values. However, it may be possible to form pseudoreplications by clustering responses where the explanatory values are approximately identical. The reader is referred to Daniel and Wood (1969), where these methods are discussed in considerable detail. A more plausible way to investigate heteroscedasticity in multiple regression is by clustering observations according to prior, natural, and meaningful associations. As an example, we shall analyze data on state education expenditures. The data has been previously mentioned in Chapter 4.

5.6. EDUCATION EXPENDITURE DATA

This data (EDUC75) consists of per capita expenditure on education, per capita personal income, proportion of population under 18 years of age and proportion of population residing in urban areas. In Chapter 4 it was suggested that this data be looked at across time (the data is available for 1965, 1970, and 1975) to check on the stability of the coefficients. For the present analysis we shall work only with the 1975 data. The objective is to get the best representation of the relationship between expenditure on education and the other variables using data for all 50 states. The data is grouped in a natural way, by geographic region. Our assumption is that although the relationship is structurally the same in each region, the coefficients and residual variances may differ from region to region. The different variances constitute a case of heteroscedasticity that can be treated directly in the analysis.

The variables and the form of the model are the same as those used by Anscombe (1975) for his analysis of the 1970 data. The variable names and definitions appear in Table 5.2.

The model is

$$SE75 = \beta_0 + \beta_1 \cdot PI73 + \beta_2 \cdot Y74 + \beta_3 \cdot URB70 + u. \qquad (5.7)$$

The data is presented in Table 5.2a. States may be grouped into geographic

Table 5.2   State expenditures on education, variable list

Variable name	Description
SE75	Per capita expenditure on education projected for 1975
PI73	Per capita income in 1973
Y74	Number of residents per thousand under 18 years of age in 1974
URB70	Number of residents per thousand living in urban areas in 1970

Table 5.2a Education expenditure data[a]

CASE	STATE	URB70	SE75	PI73	Y74	GEOG
1	ME	508	235	3944	325	1
2	NH	564	231	4578	323	1
3	VT	322	270	4011	328	1
4	MA	846	261	5233	305	1
5	RI	871	300	4780	303	1
6	CT	774	317	5889	307	1
7	NY	856	387	5663	301	1
8	NJ	889	285	5759	310	1
9	PA	715	300	4894	300	1
10	OH	753	221	5012	324	2
11	IN	649	264	4908	329	2
12	IL	830	308	5753	320	2
13	MI	738	379	5439	337	2
14	WI	659	342	4634	328	2
15	MN	664	378	4921	330	2
16	IA	572	232	4869	318	2
17	MO	701	231	4672	309	2
18	ND	443	246	4782	333	2
19	SD	446	230	4296	330	2
20	NB	615	268	4827	318	2
21	KS	661	337	5057	304	2
22	DE	722	344	5540	328	3
23	MD	766	330	5331	323	3
24	VA	631	261	4715	317	3
25	WV	390	214	3828	310	3
26	NC	450	245	4120	321	3
27	SC	476	233	3817	342	3
28	GA	603	250	4243	339	3
29	FL	805	243	4647	287	3
30	KY	523	216	3967	325	3
31	TN	588	212	3946	315	3
32	AL	584	208	3724	332	3
33	MS	445	215	3448	358	3
34	AR	500	221	3680	320	3
35	LA	661	244	3825	355	3
36	OK	680	234	4189	306	3
37	TX	797	269	4336	335	3
38	MT	534	302	4418	335	4
39	ID	541	268	4323	344	4
40	WY	605	323	4813	331	4
41	CO	785	304	5046	324	4
42	NM	698	317	3764	366	4
43	AZ	796	332	4504	340	4
44	UT	804	315	4005	378	4
45	NV	809	291	5560	330	4
46	WA	726	312	4989	313	4
47	OR	671	316	4697	305	4
48	CA	909	332	5438	307	4
49	AK	484	546	5613	386	4
50	HI	831	311	5309	333	4

[a]The data has been altered for illustrative purposes.

regions based on the presumption that there exists a sense of regional homogeneity. The four broad geographic regions (GEOG), (1) northeast, (2) north central, (3) south, and (4) west, are used to define the groups. It should be noted that data could be analyzed using dummy variables to look for special effects associated with the regions or to formulate tests for the equality of regressions across regions. However, our objective here is to develop one relationship that can serve as the best representation for all regions and all states. This goal is accomplished by taking regional differences into account through an extension of the method of weighted least squares.

It is assumed that there is a unique residual variance associated with each of the four regions. The variances are denoted as $(c_1\sigma)^2$, $(c_2\sigma)^2$, $(c_3\sigma)^2$, and $(c_4\sigma)^2$, where σ is the common part and the c_i's are unique to the regions. According to the principle of weighted least squares, the regression coefficients should be determined by minimizing

$$S_w = S_1 + S_2 + S_3 + S_4,$$

where

$$S_i = \sum_{j=1}^{n_i} \frac{1}{c_i^2}(\text{SE75}_j - \beta_0 - \beta_1 \cdot \text{PI73}_j - \beta_2 \cdot \text{Y74}_j - \beta_3 \cdot \text{URB70}_j)^2$$

$$i = 1, 2, 3, 4.$$

Each of S_1 through S_4 corresponds to a region, and the sum is taken over only those states that are in the region. The factors $\{1/c_i^2\}$ are the weights that determine how much influence each residual has in estimating the regression coefficients. The weighting scheme is intuitively justified by arguing that observations that are most erratic (large residual variance) should have little influence in determining the coefficients.

The weighted least squares estimates can also be justified by a second argument. The object is to transform the data so that the parameters of the model are unaffected, but the residual variance in the transformed model is constant. The prescribed transformation is to divide each observation by the appropriate c_i resulting in a regression of $\text{SE75}/c_i$ versus $1/c_i$, $\text{PI73}/c_i$, $\text{Y74}/c_i$, and $\text{URB70}/c_i$.* Then, the residual term, in concept, is also divided by c_i, the resulting residuals have a common variance, σ^2, and the estimated coefficients have all the standard least square properties.

---

*If we denote a variable with a double subscript, i and j, with i representing region and j representing observation within region, then each variable for an observation in region i is divided by c_i. Note that $1/c_i$ is the transformed variable corresponding to β_0. The transformed model is

$$\frac{\text{SE75}_{ij}}{c_i} = \beta_0 \frac{1}{c_i} + \beta_1 \frac{\text{PI73}_{ij}}{c_i} + \beta_2 \frac{\text{Y74}_{ij}}{c_i} + \beta_3 \frac{\text{URB70}_{ij}}{c_i} + u'_{ij}$$

and the variance of u'_{ij} is σ^2. Notice that the same regression coefficients appear in the transformed model as in the original model. Also, the intercept is zero in the transformed model.

The values of the c_i's are unknown and must be estimated in the same sense that σ^2 and the β's must be estimated. We propose a two-stage estimation procedure. In the first stage perform a regression using the raw data as prescribed in the model of Equation (5.7). Use the empirical residuals grouped by region to compute an estimate of regional residual variance. For example, in the northeast, compute $s_1^2 = \sum e_i^2/9$, where the sum is taken over the nine residuals corresponding to the nine states in the northeast. Compute s_2^2, s_3^2, and s_4^2 in a similar fashion, and obtain s^2 as the weighted average of s_1^2, s_2^2, s_3^2, and s_4^2. Then c_i is estimated as $(s_i^2/s^2)^{1/2}$. The regression results for step 1 using data from all 50 states are in Table 5.3. Two residual plots were prepared to check on specification. The residuals were plotted against a categorical variable designating region (Figure 5.3) and against the fitted values (Figure 5.4). Figure 5.3 suggests that the spread of the residuals is different for the different regions; that is that the variances are not equal. The purpose of Figure 5.4 is to look for patterns in the size and variation of the residuals as a function of the fitted values of y. The observed scatter of points does not give any indication of heteroscedasticity beyond what was learned in Figure 5.3. However, the residual for Alaska appears to be an outlier. The size of the residual is not seriously out of range, but the value of fitted expenditure is quite large relative to the other states. Checking the raw data shows that SE75 for Alaska is over \$150 larger than SE75 for any other state, but the values of the explanatory variables are not as extreme. The data for Alaska may have an undue influence on determining the regression coefficients. To check this possibility, the regression was recomputed with Alaska excluded. The estimated values of the coefficients changed significantly. See Table 5.4. Plots similar to those of Figures 5.3 and 5.4 are presented as Figures 5.5 and 5.6. With Alaska removed, Figure 5.5 does not give as strong an indication of heteroscedasticity as was found in Figure 5.3. Nevertheless, the observation is excluded for the remainder of the analysis because it represents a special situation that has too much influence on the regression results. Note that Alaska's influence would not be diminished

Table 5.3   Regression results: State expenditures on
education ($n = 50$)

Variable	Coefficient	SE	t
PI73	.072	.012	6.00
Y74	1.554	.315	4.93
URB70	− .004	.051	− 0.08
CONSTANT	− 556.900	123.200	− 4.52
$R^2 = .591$		$s = 40.462$	

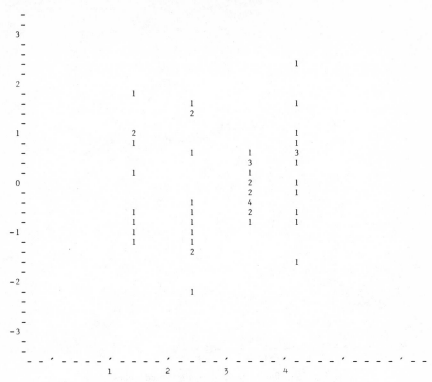

Fig. 5.3. Plot of standardized residuals against regions.

by the weighting scheme because its residual value is not particularly large.

To proceed with the analysis we must obtain the weights. They are computed from the OLS residuals by the method described above and appear in Table 5.5. The WLS regression results appear in Table 5.6 along with the OLS results for comparison. Residuals from the transformed model are plotted in Figures 5.7 and 5.8. From Figure 5.7 it appears that the spread of residuals by geographic region has evened out compared to

Table 5.4   Regression results: State expenditures
on education ($n = 49$) Alaska omitted

Variable	Coefficient	SE	t
PI73	.048	.012	4.00
Y74	.889	.331	2.69
URB70	.067	.049	1.37
CONSTANT	− 278.000	132.390	2.10
$R^2 = .497$	$s = 35.808$		

Fig. 5.4. Plot of residuals against fitted values.

the previous figures. Also there is no pattern in the plot of the residuals against the fitted values of Y (Figure 5.8). The WLS solution is preferred to the OLS solution. Referring to Table 5.6, we see that the WLS solution does not fit the historical data as well as the OLS solution when considering s or R^2 as indicators of goodness of fit.* This result is expected since one of the important properties of OLS is that it provides a solution with minimum s or, equivalently, maximum R^2. Our choice of the WLS solution is based on the pattern of the residuals.

*Note that for comparative purposes, s for the WLS solution is computed as the square root of

$$s^2 = \sum (\text{SE75}_i - \widehat{\text{SE75}}_i)^2 / 45,$$

where the summation is over all observations and

$$\widehat{\text{SE75}}_i = -320.1 + .063 \cdot \text{PI73}_i + .877 \cdot \text{Y74}_i + .027 \cdot \text{URB70}_i.$$

That is, $\widehat{\text{SE75}}_i$ is computed in terms of the WLS estimated coefficients and the weights, $\{c_i\}$, play no further role in the computation of s.

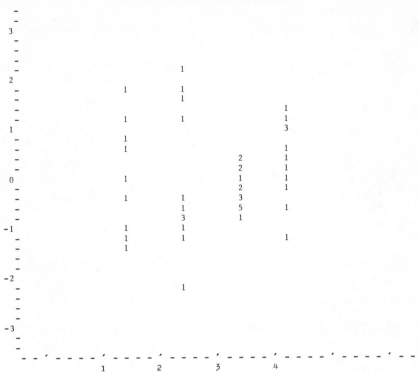

Fig. 5.5.   Plot of residuals against regions (excluding Alaska).

The differences in the solutions are not dramatic. In both solutions, PI73 is the most important explanatory variable followed by Y74 and URB70. One noticeable difference is that the standard error of the coefficient of Y74 in the WLS solution is larger than it was in the OLS solution. In fact it is large enough so that Y74 no longer seems to make a significant contribution to explaining the variation in SE75 ($t = 1.282$). It is not possible to make a precise test of significance because exact distribution theory for the two-stage procedure used to obtain the WLS solution has not been worked out. If the weights were known in advance rather than as estimates from data, then the statistical tests based on the WLS procedure would be exact. Of course, it is difficult to imagine a situation similar to the one being discussed where the weights would be known in advance. Nevertheless, based on the empirical analysis above, there is a clear suggestion that weighting is required and that neither of the population variables, Y74 or URB70 are important factors. In addition, since less than 50% of the variation in SE75 has been explained ($R^2 = .471$), the search for other factors must continue.

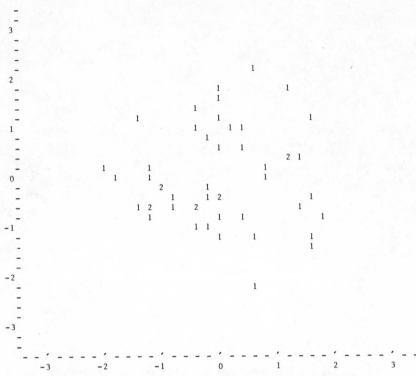

Fig. 5.6.   Plot of residuals against fitted values (excluding Alaska).

Table 5.5   Weights for weighted least squares

	n	c
Northeast	9	1.110
North Central	12	1.433
South	16	.383
West	12	.794

Table 5.6   OLS and WLS coefficients for education data (n = 49) Alaska omitted

	OLS			WLS		
Variable	Coefficient	SE	t	Coefficient	SE	t
PI73	.048	.012	4.000	.063	.007	9.000
Y74	.889	.331	2.685	.877	.684	1.282
URB70	.067	.049	1.367	.027	.032	.843
CONSTANT	− 278.000			− 320.100		

$R^2 = .497$ $s = 35.808$ $R^2 = .471$ $s = 36.711$

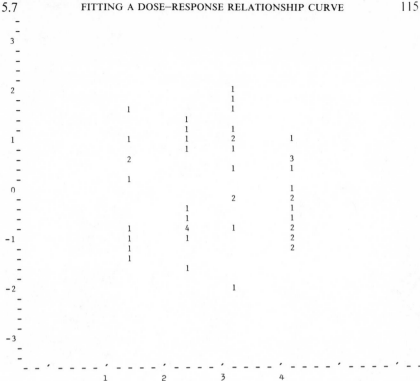

Fig. 5.7. Standardized residuals by geographic region for WLS solution.

5.7. FITTING A DOSE–RESPONSE RELATIONSHIP CURVE

An important area for the application of weighted least square analysis is the fitting of a linear regression line when the dependent variable y is a proportion. Consider the following situation: An experimenter can administer a stimulus at different levels. Subjects are assigned at random to different levels of the stimulus and for each subject a binary response is noted. From this set of observations, a relationship between the stimulus and the proportion responding to the stimulus is constructed. A very common example is in the field of pharmacology, in bioassay, where the levels of stimulus may represent different doses of a drug or poison, and the binary response is death or survival. Another example is in the field of consumer behavior study where the stimulus is the discounts offered and the binary response is the purchase or nonpurchase of some merchandise.

Suppose a pesticide is tried at k different levels. At the ith level of dosage x_i, let r_i be the number of insects dying out of a total n_i exposed to the pesticide ($i = 1, 2, \ldots, k$). We want to estimate the relationship between dose x_i and the proportion dying $p_i = r_i / n_i$. The sample proportion p_i is a

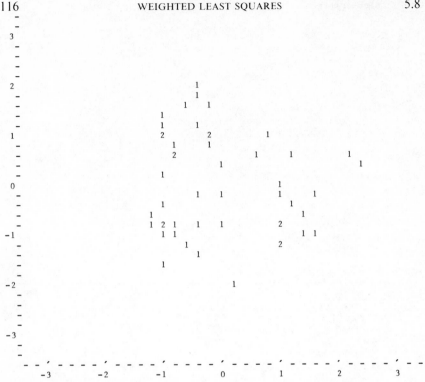

Fig. 5.8. Standardized residuals versus fitted values for WLS solution.

binomial variate, with an expected value P_i, and variance $P_i(1 - P_i)/n_i$ where P_i is the population probability of death for a subject who receives a dose x_i. The relationship between p_i and x_i is based on the notion that

$$P_i = f(x_i), \tag{5.8}$$

where the function $f(\cdot)$ is increasing (or at least not decreasing) with x, and is bounded between 0 and 1. The function should satisfy these properties because (i) P being a probability is bounded between 0 and 1 and (ii) if the pesticide is toxic, higher doses should decrease the chances of survival for a subject. These considerations effectively rule out the linear model

$$P_i = \alpha + \beta x_i \tag{5.9}$$

because for the above model, P_i is not bounded.

5.8.   THE LOGISTIC MODEL

The stimulus response relationships have generally been found to be nonlinear. A nonlinear function which has been found to accurately represent the relationship between dose x_i and the proportion dying is

$$P_i = \frac{e^{\beta_0 + \beta_1 x_i}}{1 + e^{\beta_0 + \beta_1 x_i}}.$$ (5.10)

The relationship (5.10) is called the logistic response function and has the shape given in Figure 5.9. It is seen that the logistic function is bounded between 0 and 1, and is monotonic. Physical considerations based on concepts of threshold values provide a heuristic justification for the use of (5.10) to represent a stimulus–response relationship (Cox, 1970).

The setup described above differs considerably from those of our other examples. In the present situation the experimenter has the control of dosages or stimuli, and can use replication to estimate the variability of response at each dose level. This is a designed, experimental study, unlike the others which were observational.

The objectives for this type of analysis are not only to determine the nature of dose–response relationship but also to estimate the dosages which induce specified levels of response. Of particular interest is the dosage which produces a response in 50% of the population (median dose).

The logistic model (this application is sometimes called logit analysis) has been used extensively in biological work. For analyzing proportions from binary response data, it is a very appealing model and easy to fit. An

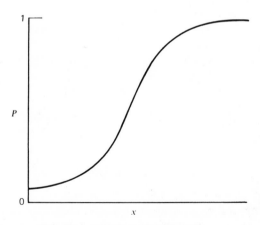

Fig. 5.9.   Logistic response function.

alternative model in which the response function is represented by the cumulative distribution function of the normal probability distribution has also been proposed. The cumulative curve of the normal distribution has a shape similar to that of the logistic function. The second model is called the probit-model, and for details we refer the reader to Finney (1964). Besides medicine and pharmacology, the logistic model has been used in learning theory, in the study of consumer behavior (response to advertising) and market promotion studies.

5.9. FITTING A LOGISTIC RESPONSE FUNCTION

Since the response function (logistic) to be fitted is nonlinear, we will work with transformed variables. The transformation is chosen so that the response function is made linear. Further, since the variables have nonconstant variance, for a correct analysis we must use weighted least square methods on the transformed variables.

From (5.10) it is seen that the relation between P_i, the expected response, and dose x_i is not linear. The tranformation

$$P_i' = \ln \frac{P_i}{1 - P_i} \qquad (5.11)$$

is, however, seen to reduce (5.10) to

$$P_i' = \beta_0 + \beta_1 x_i. \qquad (5.12)$$

To arrive at a relationship between dose x_i and the observed proportion p_i, we convert p_i to a new variable p_i'

$$p_i' = \ln \frac{p_i}{1 - p_i} \qquad (5.13)$$

and use the model

$$p_i' = \beta_0 + \beta_1 x_i + u_i. \qquad (5.14)$$

The transformed p_i' has mean approximately equal to $\ln(P_i/(1 - P_i))$ and variance of σ_i^2

$$\sigma_i^2 = \frac{1}{n_i P_i (1 - P_i)}. \qquad (5.15)$$

The regression of p_i' on x_i is clearly heteroscedastic, and it should not be estimated using ordinary least square methods. An estimate of σ_i^2 is obtained by replacing the P_i in equation by their sample estimates p_i.

To fit (5.14), we therefore use a weighted least square method, the weight attached to the ith point being

$$w_i = n_i p_i (1 - p_i).$$ (5.16)

The estimates b_0 and b_1 of β_0 and β_1 are

$$b_1 = \frac{\sum_{i}^{k} w_i (p_i' - \bar{p}')(x_i - \bar{x})}{\sum w_i (x_i - \bar{x})^2},$$ (5.17)

$$b_0 = \bar{p}' - b_1 \bar{x},$$ (5.18)

where

$$\bar{p}' = \frac{\sum w_i p_i'}{\sum w_i} \quad \text{and} \quad \bar{x} = \frac{\sum w_i x_i}{\sum w_i}.$$

As estimates of variances of b_1 and b_0 we take $s^2(b_1)$ and $s^2(b_0)$ where

$$s^2(b_1) = \frac{1}{\sum w_i (x_i - \bar{x})^2}$$ (5.19)

and

$$s^2(b_0) = \frac{1}{\sum w_i} + \bar{x}^2 s^2(b_1).$$ (5.20)

Expressions (5.19) and (5.20) provide accurate estimates of the variance when the number of subjects treated at each dose x_i is large, the response relationship is logistic, and levels of the dose x_i are fixed (nonstochastic).

The value of x at $P = 0.5$ is of considerable practical importance. It is called the median dose and is denoted by x_{50}. The median dose is the level of the stimulus which produces a response in 50% of the population. The value of $P = 0.5$ corresponds to a value of $P' = 0$ as can be easily seen from (5.11). Hence, $x_{50} = -\beta_0/\beta_1$. As an estimate of the median dose $\hat{x}_{50}$ we take $-b_0/b_1$. In most bioassay work x is usually taken as the logarithm of the dosage, so to get $\hat{D}_{50}$ we must use $10^{\hat{x}_{50}}$. The estimated variances of the median doses are

$$s_{\hat{x}_{50}}^2 = \frac{1}{b_1^2}\left[\frac{1}{\sum w_i} + s^2(b_1)(\hat{x}_{50} - \bar{x})^2\right],$$

where x, the dose, is measured directly, or

$$s_{\hat{D}_{50}}^2 = s_{\hat{x}_{50}}^2 \left(\frac{\hat{D}_{50}}{\log e} \right)^2,$$

where

$$\hat{x}_{50} = \log \hat{D}_{50};$$

that is, x is measured on a logarithmic scale.

From (5.13) it is seen that the logit transformation is not defined for $p_i = 0$ or 1, that is, for those levels of dosage where none or all of the subjects die. For fitting a dose–response relationship, such points are usually omitted, as there is a considerable degree of uncertainty about the exact dosage at these response levels. Does the observed value of x correspond strictly to a response $p = 0$, or could it be further increased and the response still be zero? Similar remarks apply to the top end of the scale; that is, could x be reduced and p still be 1? At the extreme range of responses the response curve is very flat. If the extreme data points are to be included, then we recommend the rule proposed by Berkson (1953) that states: Take

$$p_i = \frac{1}{2n_i} \qquad \text{if } r_i = 0$$

and

$$= 1 - \frac{1}{2n_i} \qquad \text{if } r_i = n_i.$$

Another rule which has considerable merit is to fit the model with and without the extreme data points. If the fitted models differ considerably, data points at the end of the scale should be examined. Additional experimentation may be necessary. As in any data-fitting endeavor, before adopting Equation (5.14) to describe a given data set, the adequacy of the model should be examined by looking at residual plots, examining standard errors, and so on. The fitting of a logistic response function is illustrated in the following example.

5.10.   TOXICITY OF ROTENONE

Table 5.7 presents the results of spraying rotenone of different concentrations on the chrysanthemum aphis in batches of about fifty. The number and the percentage killed at each dosage is shown. The concentra-

Table 5.7   Toxicity study of rotenone on chrysanthemum aphis

Concentration (mg/l)	$x = \log D$	Deaths r	Total n	Proportion p	p'	w
2.6	0.4150	6	50	0.120	−1.9924	5.2800
3.8	0.5797	16	48	0.333	−0.6946	10.6608
5.1	0.7076	24	46	0.522	0.0881	11.4770
7.7	0.8865	42	49	0.857	1.7906	6.0074
10.2	1.0086	44	50	0.880	1.9924	5.2800

$$p = \frac{r}{n}, \quad p' = \ln \frac{p}{1-p}, \quad w = np(1-p)$$

Source: Finney (1964).

tion D is in milligrams per liter of the solution, the dosage x is measured on the logarithmic scale $x = \log D$. The data is presented in Finney (1964) and analyzed by Berkson (1953) applying the logistic transformation.

A plot of x against p shows the observed response function to be S-shaped as in Figure 5.9. A simple straight line fit of

$$p = a + b \log x$$

to the data is unsatisfactory. Even with only five data points, a plot of the residuals against $\log x$ show a systematic pattern of variation; that is, the two end observations have negative residuals, and the three interior observations have increasing positive residuals. Based on these indications we proceed to fit a logistic response function. For calculation of b_0, b_1 and the estimates of standard errors we need

$$\sum w = 38.7052, \quad \sum wx = 27.1427, \quad \sum wx^2 = 20.3300,$$
$$\sum wp' = 4.3562, \quad \sum wpx = 12.1873.$$

The estimates of b_0 and b_1 from (5.17) and (5.18) are

$$b_1 = \frac{\sum w_i (p_i' - \bar{p}')(x_i - \bar{x})}{\sum w_i (x_i - \bar{x})^2} = 7.0485,$$

$$b_0 = \bar{p}' - b_1 \bar{x} = -4.8303.$$

The estimates of standard errors of b_0 and b_1 from (5.19) and (5.20) are

$$s(b_0) = 0.6367, \quad s(b_1) = 0.8785.$$

The estimated coefficients are highly significant, and the model gives an excellent fit to the observed data.

The estimate of the median dose on the logarithmic scale $\hat{x}_{50}$ and on the concentrations (D) scale $\hat{D}_{50}$ are

$$\hat{x}_{50} = 0.6853, \qquad \hat{D}_{50} = 4.845.$$

The estimates of their corresponding standard errors are 0.0229 and 0.2554, respectively. From the standard errors, a confidence interval for median dose can easily be constructed. The confidence intervals are first constructed on the log scale (x) and then transformed back to the dosage scale because the normality assumption is made for the log scale.

We conclude this analysis by pointing out that the logistic response function can be extended to include several variables or powers of variables. Because of the simplicity and ease of fitting, the logistic function is especially well suited for analyzing dose–response types of relationships.

REFERENCES

Anscombe, F. J., Personal communication, 1975.

Berkson, J., A statistically precise and relatively simple method of estimating the bioassay with a quantal response, based on the logistic function, *J. Amer. Stat. Assoc.*, **48**, 565–599 (1953).

Cox, D. R., *The Analysis of Binary Data*, Methuen, London, 1970.

Daniel, C. and F. S. Wood, *Fitting Equations to Data*, Wiley, New York, 1971.

Draper, N. R. and H. Smith, *Applied Regression Analysis*, Wiley, New York, 1966.

Finney, D. J., *Probit Analysis*, Cambridge University Press, London, 1964.

CHAPTER 6

The Problem of Correlated Errors

6.1. INTRODUCTION: AUTOCORRELATION

One of the standard assumptions in the regression model is that the error terms u_i and u_j, associated with the ith and jth observations, are uncorrelated. Correlation in the error terms suggests that there is additional explanatory information in the data that has not been exploited in the current model. When the observations have a *natural* sequential order, the correlation is referred to as autocorrelation.

Autocorrelation may occur for several reasons. Adjacent residuals tend to be similar in both temporal and spatial dimensions. Successive residuals in economic time series tend to be positively correlated. Large positive errors are followed by other positive errors, and large negative errors are followed by other negative errors. Observations sampled from adjacent experimental plots or areas tend to have residuals that are correlated since they are affected by similar external conditions.

The symptoms of autocorrelation may also appear as the result of a variable having been omitted from the right-hand side of the regression equation. If successive values of the omitted variable are correlated, the errors from the estimated model will appear to be correlated. When the variable is added to the equation, the apparent problem of autocorrelation can be completely eliminated.

The presence of autocorrelation has several effects on the analysis. These are summarized as follows:

1. Least square estimates are unbiased but are not efficient in the sense that they no longer have minimum variance.
2. The estimate of σ^2 and the standard errors of the regression coefficients may be seriously understated; that is, from the data the estimated standard errors would be much smaller than they actually are, giving a spurious impression of accuracy.

3. The confidence intervals and the various tests of significance commonly employed would no longer be strictly valid.

The presence of autocorrelation can be a problem of serious concern for the preceding reasons and should not be ignored.

We distinguish between two types of autocorrelation and describe methods for dealing with each. The first type is only autocorrelation in appearance. It is due to the omission of a variable that should be in the model. Once this variable is uncovered, the autocorrelation problem is resolved. The second type of autocorrelation may be referred to as pure autocorrelation. The methods of correcting for pure autocorrelation involve a transformation of the data. Formal derivations of the methods can be found in Johnston (1972) and Kmenta (1971).

6.2. CONSUMER EXPENDITURE AND MONEY STOCK

Table 6.1 gives quarterly data from 1952 to 1956 on consumer expenditure (y) and the stock of money (x) both measured in billions of current dollars for the United States. A simplified version of the quantity theory of money suggests a model given by

$$y_t = \alpha + \beta x_t + u_t,$$

where α and β are constants, u_t the error term. Economists are interested in estimating β and its standard error; β is called the multiplier and has crucial importance as an instrument in fiscal and monetary policy. Since

Table 6.1. Consumer expenditure and money stock

Year	Quarter	Consumer expenditure	Money stock	Year	Quarter	Consumer expenditure	Money stock
1952	1	214.6	159.3	1954	3	238.7	173.9
	2	217.7	161.2		4	243.2	176.1
	3	219.6	162.8	1955	1	249.4	178.0
	4	227.2	164.6		2	254.3	179.1
1953	1	230.9	165.9		3	260.9	180.2
	2	233.3	167.9		4	263.3	181.2
	3	234.1	168.3	1956	1	265.6	181.6
	4	232.3	169.7		2	268.2	182.5
1954	1	233.7	170.5		3	270.4	183.3
	2	236.5	171.6		4	275.6	184.3

Source: Milton Friedman and David Meiselman, "The relative stability of monetary velocity and the investment multiplier in the United States, 1897–1958," in *Commission on Money and Credit, Stabilization Policies*, Prentice-Hall, Englewood Cliffs, New Jersey, 1963, p. 266.

the observations are ordered in time, it is reasonable to expect that autocorrelation may be present. A summary of the regression results is given in Table 6.2.

The coefficients are significant; the standard error of the regression coefficient is .115. For a unit change in the money supply the 95% confidence interval for the change in the aggregate consumer expenditure would be (2.06, 2.54). The value of R^2 indicates that roughly 95% of the variation in the consumer expenditure can be explained by the variation in money stock. The analysis would be complete if the basic regression assumptions were valid. To check on the model assumption we examine the residuals. If there are indications that autocorrelation is present, then the model should be reestimated after eliminating the autocorrelation.

For time series data the most meaningful plot for analysis is the plot of residuals against time. The graph is given in Figure 6.1. The pattern of residuals is revealing and is characteristic of situations where the errors are correlated. Residuals of the same sign occur in clusters or bunches. The characteristic pattern would be that several successive residuals would be positive, the next several negative, and so on. From Figure 6.1 we see that the first seven residuals are positive, the next seven negative, and the last six positive. This pattern suggests that the error terms in the model are correlated and some additional analysis is required. Besides this graphical analysis, autocorrelated errors can also be detected by the Durbin–Watson statistic.

6.3. DURBIN–WATSON STATISTIC

The Durbin–Watson statistic d is the basis of a very popular test of autocorrelation in regression analysis. The test is based on the assumption that the errors constitute a first-order autoregressive series, namely

$$u_t = \rho u_{t-1} + \epsilon_t, \qquad |\rho| < 1, \qquad (6.1)$$

where ϵ_t is normally independently distributed with zero mean and constant variance. In most situations the error u_t may have a much more complex correlation structure. The first-order dependency structure, given in (6.1) is taken as a simple approximation to the actual error structure.

Table 6.2. Consumer expenditure versus money stock

Variable	Coefficient	SE	t
x	2.300	.115	20.080
CONSTANT	−154.700	19.850	−7.790
$n = 20$	$R^2 = .955$	$s = 3.983$	

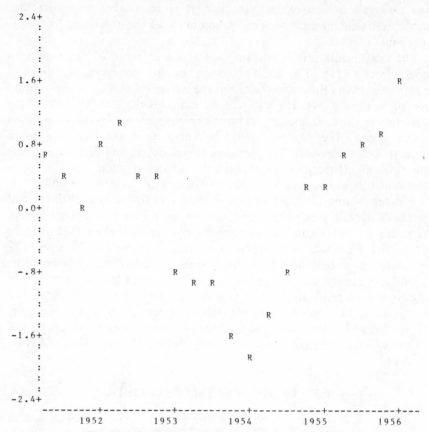

Fig. 6.1.   Graph of standardized residuals versus time.

The statistic d is defined as

$$d = \frac{\sum\limits_{2}^{n} (e_t - e_{t-1})^2}{\sum\limits_{2}^{n} e_t^2}$$

and is used for testing the null hypothesis $H_0(\rho = 0)$ against an alternative $H_1(\rho > 0)$. Note that when $\rho = 0$ in Equation (6.1), the u's are uncorrelated.

We estimate the parameter ρ by r where

$$r = \frac{\sum\limits_{2}^{n} e_t e_{t-1}}{\sum\limits_{2}^{n} e_t^2}.$$

An approximate relationship exists between d and r,

$$d \doteq 2(1 - r),$$

showing that d has a range of 0 to 4. Since r is an estimate of ρ, it is clear that d is close to 2 when $\rho = 0$ and near to zero when $\rho = 1$. The closer the sample value of d to 2, the firmer the evidence that there is no autocorrelation present in the error. Evidence of autocorelation is indicated by the deviation of d from the numerical value of 2. The formal test for autocorrelation operates as follows: Calculate the sample statistic d. If

   (i) $d < d_L$ reject H_0
   (ii) $d > d_U$ do not reject H_0,
   (iii) $d_L < d < d_U$ the test is inconclusive.

The values of (d_L, d_U) for different percentage points have been tabulated by Durbin and Watson. A table is found in the Appendix.

   Tests for negative autocorrelation are seldom performed; situations where negative autocorrelation is likely to occur are difficult to envisage and are rare. If, however, a test is desired then instead of working with d, one works with $(4 - d)$, and follows the same procedure as for the testing of positive autocorrelation.

   In our money stock and consumer expenditure data the value of d is 0.328. From the table in the Appendix, $d_L = 1.18$ and we find that the value of d is significant at the 5% level. H_0 is rejected, thereby showing that there is autocorrelation present. This essentially reconfirms our earlier conclusion which was arrived at by looking at the residual plots against time.

   If d had been larger than $d_u = 1.4$, autocorrelation would not be a problem and no further analysis is needed. When $d_L < d < d_U$, additional analysis of the equation is optional. We suggest that in cases where the Durbin–Watson statistic lies in the inconclusive region, reestimate the equation using the methods described below to see if any major changes occur.

   As pointed out earlier, the presence of correlated errors distorts estimates of standard errors, confidence intervals, and statistical tests, and, therefore, we should reestimate the equation. When autocorrelated errors are indicated, two approaches may be followed. These are: (i) work with transformed variables or (ii) introduce additional variables which have in them time-ordered effects. We illustrate the first approach with the money stock data. The second approach is illustrated in section 6.6.

6.4. REMOVAL OF AUTOCORRELATION BY TRANSFORMATION

When the residual plots and Durbin–Watson statistic indicate the presence of correlated errors the estimated regression equation should be refitted taking the autocorrelation into account. One method for adjusting the model is the use of a transformation that involves the unknown autocorrelation parameter, ρ. The introduction of ρ causes the model to be nonlinear. The direct application of least squares is not possible. However, there are a number of procedures that may be used to circumvent the nonlinearity (Johnston, 1972). We shall use the method due to Cochrane and Orcutt (1949).

When the errors have an autoregressive structure as given in (6.1), it is seen that by transforming to $(y_t - \rho y_{t-1})$ and $(x_t - \rho x_{t-1})$ the errors in the model are uncorrelated. We can see this by substituting the transformed variable in the original model.

$$y_t - \rho y_{t-1} = \alpha + \beta x_t - \alpha\rho - \beta\rho x_{t-1} + u_t - \rho u_{t-1}$$

$$= \alpha(1-\rho) + \beta(x_t - \rho x_{t-1}) + \epsilon_t, \tag{6.2}$$

$$y_t^* = \alpha^* + \beta^* x_t^* + \epsilon_t, \tag{6.3}$$

where $y_t^* = y_t - \rho y_{t-1}$; $x_t^* = x_t - \rho x_{t-1}$; $\alpha^* = \alpha(1-\rho)$, $\beta^* = \beta$. The ϵ_t's are uncorrelated and satisfy the assumptions of the standard linear model. The value of ρ is unknown and has to be estimated from the data. Cochrane –Orcutt have proposed an iterative procedure. The procedure operates as follows:

1. Compute the ordinary least square estimates of α and β.
2. Calculate the residuals and from the residuals estimate ρ by

$$\hat{\rho} = \frac{\sum_{2}^{n} e_t e_{t-1}}{\sum_{2}^{n} e_{t-1}^2}.$$

3. Fit the equation given in (6.3) using variables $(y_t - \hat{\rho} y_{t-1})$ and

$(x_t - \hat{\rho} x_{t-1})$. The estimates of the parameters in the original equations are $\hat{\alpha} = \hat{\alpha}^* / (1 - \hat{\rho})$ and $\hat{\beta} = \hat{\beta}^*$.

4. Examine the residuals of the newly fitted equation. If the new residuals show no autocorrelation terminate the procedure, otherwise continue the whole procedure using the estimates $\hat{\alpha}$ and $\hat{\beta}$ as estimates of α and β instead of the original least square estimates.

As a practical rule we suggest that if the first application of Cochrane–Orcutt procedure does not yield nonautocorrelated residuals, then one should look for alternative methods of removing autocorrelation. We apply the Cochrane–Orcutt procedure to the data given in Table 6.1.

The d value for the original data is 0.328, which is highly significant. The value of $\hat{\rho}$ is .874. On fitting the regression equation to the variables $(y_t - .874 y_{t-1})$ and $(x_t - .874 x_{t-1})$ we have a d value of 1.607. The value of d_u for $n = 19$ and $K = 1$ (the number of variables in the equation) is 1.40 at the 5% level. Consequently $H_0(\rho = 0)$ is not rejected*. The fitted equation is

$$y_t = -324.44 + 2.758 x_t$$

with a standard error for b of 0.444 as opposed to the least square estimate of the original equation which was $y_t = -154.7 + 2.300 x_t$ with a standard error for b of 0.115. The newly estimated standard error is larger by a factor of almost 4. The residual plots for the fitted equation of the transformed variables are shown in Figure 6.2. The residual plots show less clustering of the adjacent residuals by sign and the Cochrane–Orcutt procedure has worked to our advantage.

6.5. ITERATIVE ESTIMATION WITH AUTOCORRELATED ERRORS

One advantage of the Cochrane–Orcutt procedure is that estimates of the parameters are obtained using standard least squares computations. Although two stages are required, the procedure is relatively simple. A more direct approach is to try to estimate values of ρ, α, and β simultaneously. The model is formulated as before requiring the construction of transformed variables $y_t - \rho y_{t-1}$ and $x_t - \rho x_{t-1}$. Parameter estimates are obtained by minimizing the sum of squared errors which is given as

$$S(\alpha, \beta, \rho) = \sum_{t=2}^{n} \left[y_t - \rho y_{t-1} - \alpha(1 - \rho) - \beta(x_t - \rho x_{t-1}) \right]^2.$$

*The significance level of the test is not exact because $\hat{\rho}$ was used in the estimation process. The d value of 1.607 may be viewed as an index of autocorrelation that indicates an improvement from the previous value of .328.

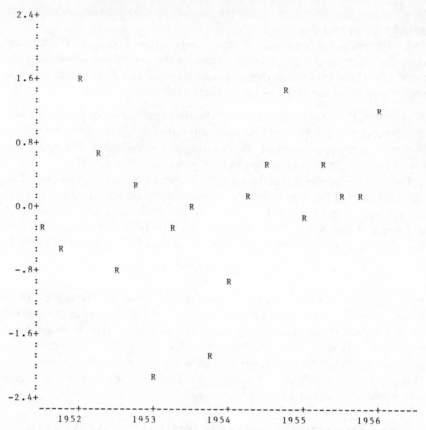

Fig. 6.2. Graph of standardized residuals versus time after one iteration of Cochrane–Orcutt
method.

If the value of ρ were known, then α and β are easily obtained by
regressing $y_t - \rho y_{t-1}$ on $x_t - \rho x_{t-1}$. Final estimates are obtained by searching through many values of ρ until a combination of ρ, α, and β are found
that minimizes $S(\rho, \alpha, \beta)$. The search could be accomplished using a standard regression computer program, but the process can be much more
efficient with an automated search procedure. This method is due to
Hildreth and Lu. For a discussion of the estimation procedure and
properties of the estimates obtained, see Kmenta (1971).
    Once the minimizing values, say $\tilde{\rho}$, $\tilde{\alpha}$, and $\tilde{\beta}$, have been obtained, the
standard error for the estimate of β can be approximated using a version
of Equation (1.4) of Chapter 1. The formula is used as though $y_t - \rho y_{t-1}$

were regressed on $x_t - \rho x_{t-1}$ with ρ known; that is, the estimated standard error of $\tilde{\beta}$ is

$$\text{s.e.}(\tilde{\beta}) = \frac{s}{\left\{ \sum \left[x_t - \tilde{\rho} x_{t-1} - \bar{x}(1-\tilde{\rho}) \right]^2 \right\}^{1/2}}$$

and s is the square root of $S(\tilde{\rho}, \tilde{\alpha}, \tilde{\beta})/(n-2)$. When adequate computing facilities are available such that the iterative computations are easy to accomplish, then the latter method is recommended. However, it is not expected that the estimates and standard errors for the iterative method and the two-stage Cochrane–Orcutt method would be appreciably different. The estimates from the three methods, ordinary least squares, Cochrane–Orcutt, and iterative for the data of Table 6.1, are given in Table 6.2a for comparison.

6.6.   AUTOCORRELATION AND MISSING VARIABLES

The characteristics of the regression residuals that suggest autocorrelation may also be indicative of other aspects of faulty model specification. In the previous example, the plot of residuals against time and the statistical test based on the Durbin–Watson statistic were used to conclude that the residuals were autocorrelated. Autocorrelation is only one of a number of possible explanations for the clustered type of residual plot or low Durbin–Watson value.

In general, a plot of residuals against any one of the list of potential independent variables may uncover additional information that can be used to further explain variation in the dependent variable. When the residuals plotted against time show a pattern of the type described in the previous example, it is reasonable to suspect that it may be due to the omission of variables that change over time. Certainly when the residuals appear in clusters alternating above and below the mean value line of zero, when the estimated autocorrelation coefficient is large and the Durbin–

Table 6.2a.   Comparison of regression estimates

Parameter	Ordinary least squares	Cochrane–Orcutt	Iterative
ρ		.874	.824
α	-154.700	-324.44	-235.509
β	2.300	2.758	2.753
SE (β)	.115	.444	.436

Watson statistic is significant, it would appear that the presence of auto-correlation is overwhelmingly supported. We shall see that this conclusion may be incorrect. The observed symptoms would be better interpreted initially as a general indication of some form of model misspecification. All possible correction procedures should be considered. In fact, it is always better to explore fully the possibility of some additional explanatory variables before yielding to an autoregressive model for the error structure. It is more satisfying and probably more useful to be able to understand the source of apparent autocorrelation in terms of an additional variable. The marginal effect of that variable can then be estimated and used in an informative way. The transformations that correct for pure autocorrelation may be viewed as an action of last resort.

6.7. ANALYSIS OF HOUSING STARTS

As an example of a situation where autocorrelation appears artificially because of the omission of another explanatory variable, consider the following project undertaken by a midwestern construction industry association. The association wants to have a better understanding of the relationship between housing starts and population growth. They are interested in being able to forecast construction activity. Their approach is to develop annual data on regional housing starts and try to relate this to potential home buyers in the region. Realizing that it is almost impossible to measure the number of potential house buyers accurately, the researchers settled for the size of the 22 to 44-year old population group in the region as a variable that reflects the size of potential home buyers. With some diligent work they were able to bring together 25 years of historical data for the region (see Table 6.3). Their goal was to get a simple regression relationship between housing starts and population. Then using some methods that they developed for projecting population changes they would be able to estimate corresponding changes in the requirements for new houses. The construction association was aware that the relationship between population and housing starts could be very complex. It is even reasonable to suggest that housing affects population growth (by migration) instead of the other way around. Although the proposed model is undoubtedly naive it serves a useful purpose as a starting point for their analysis.

Analysis

The regression results for the 25 years of data are given in Table 6.4. The proportion of explained variation is .9252 and we see that an increase in

Table 6.3.   Data for housing starts

ROW	POPULT	STARTS	IND
* 1 *	2.200	0.09090	0.03635
* 2 *	2.222	0.08942	0.03345
* 3 *	2.244	0.09755	0.03870
* 4 *	2.267	0.09550	0.03745
* 5 *	2.280	0.09678	0.04063
* 6 *	2.289	0.10327	0.04237
* 7 *	2.289	0.10513	0.04715
* 8 *	2.290	0.10840	0.04883
* 9 *	2.299	0.10822	0.04836
* 10 *	2.300	0.10741	0.05160
* 11 *	2.300	0.10751	0.04879
* 12 *	2.340	0.11429	0.05523
* 13 *	2.386	0.11048	0.04770
* 14 *	2.433	0.11604	0.05282
* 15 *	2.482	0.11688	0.05473
* 16 *	2.532	0.12044	0.05531
* 17 *	2.580	0.12125	0.05898
* 18 *	2.605	0.12080	0.06267
* 19 *	2.631	0.12368	0.05462
* 20 *	2.658	0.12679	0.05672
* 21 *	2.684	0.12996	0.06674
* 22 *	2.711	0.13445	0.06451
* 23 *	2.738	0.13325	0.06313
* 24 *	2.766	0.13863	0.06573
* 25 *	2.793	0.13964	0.07229

Source: Generated by the authors.

population of one million leads to an increase in housing starts of about 70,000. The Durbin–Watson statistic and the residual plot (Figure 6.3) against time suggest strong autocorrelation. However, it is fairly simple to conjecture about other variables that may further explain housing starts and could be responsible for the appearance of autocorrelation. These variables include the unemployment rate, social trends in marriage and family formation, government programs in housing, and the availability of construction and mortgage funds. The first choice was an index that measures the availability of mortgage money for the region. Adding that variable to the equation the model is

$$\text{HS}_t = \beta_0 + \beta_1 (\text{POP})_t + \beta_2 (\text{IND})_t + u_t.$$

Table 6.4.   Regression on housing starts versus population

Variable	Coefficient	SE	t
POP	.0714	.0042	16.865
CONSTANT	− .0609	.0104	
$R^2 = .9252$	$d = 0.621$	$s = 0.0041$	

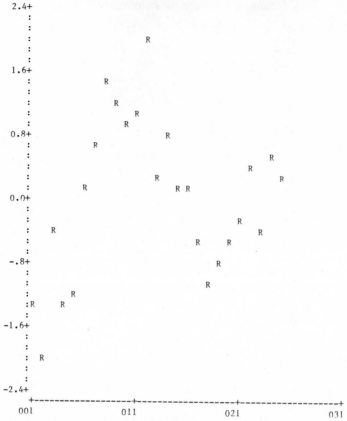

Fig. 6.3.   Graph of housing start residuals against time.

The introduction of the additional variable has the effect of removing autocorrelation. From Table 6.5 we see that the Durbin–Watson statistic has the new value 2.29, well into the acceptable region. The graph of the residuals against time (Figure 6.4) is also improved. The regression coefficients and their corresponding t values show that there is a significant population effect but that it was overstated by a factor of more than 2 in the first equation. In a certain sense the effect of changes in the availability of mortgage money for a fixed level of population is more important than a similar change in population. If each variable in the regression equation is replaced by the standardized version of the variable (the variables transformed so as to have mean 0, and unit variance) then the resulting

Table 6.5.   Regression on housing starts versus population and index

Variable	Coefficient	SE	t
POP	.0332	.0053	6.292
IND	.8147	.1009	8.070
CONSTANT	−.0095	.0084	
$R^2 = .9826$	$d = 2.290$	$s = 0.0020$	

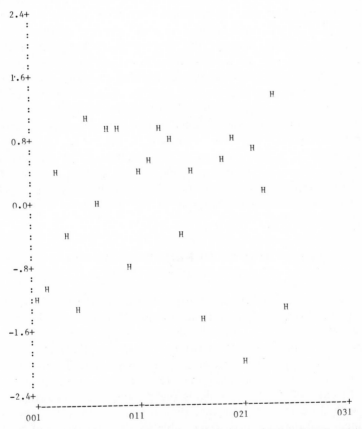

Fig. 6.4.   Graph of residuals from regression on population and index plotted against time.

regression equation is

$$\widetilde{\text{HS}}_t = .4439 \, (\widetilde{\text{POP}})_t + .5693 \, (\widetilde{\text{IND}})_t,$$

($\tilde{Y}$ denotes the standardized value of Y, $\tilde{Y} = (Y - \bar{Y})/s_Y$). A unit increase in the standardized value of POP_t is worth an additional .4439 to the standardized value of HS_t, that is if the population increases by 1 standard deviation then HS_t increases by 0.4439 standard deviation. Similarly, if IND_t increases by 1 standard deviation HS_t increases by 0.5693 standard deviation. Therefore, in terms of the standardized variables, the mortgage index is more important (has a larger effect) than population size.

The example on housing starts illustrates two important points. First, a large value of R^2 does not imply that the data has been fitted and explained well. Any pair of variables that show trends over time are usually highly correlated. A large value of R^2 does not necessarily confirm that the relationship between the two variables has been adequately characterized. Second, the Durbin–Watson statistic as well as the residual plots may indicate the presence of autocorrelation among the errors when in fact the errors are independent but the omission of a variable or variables has given rise to the observed situation. Even though the Durbin–Watson statistic was designed to detect first-order autocorrelation it can have a significant value when some other model assumptions are violated such as misspecification of the variables to be included in the model. In general, a significant value of the Durbin–Watson statistic should be interpreted as an indication that some problem exists and both the possibility of a missing variable or the presence of autocorrelation should be considered.

6.8. LIMITATIONS OF DURBIN–WATSON STATISTIC: SKI EQUIPMENT SALES

In the previous examples on expenditure versus money stock, and housing starts versus population size the residuals from the initial regression equations indicated model misspecifications associated with time dependence. In both cases the Durbin–Watson statistic was small enough to conclude that positive autocorrelation was present. The residual plots against time further confirmed the presence of a time-dependent error term. In each of the two problems the presence of autocorrelation was dealt with differently. In one case (housing starts) an additional variable was uncovered that had been responsible for the appearance of autocorrelation and in the other case (money stock) the Cochrane–Orcutt method was used to deal with what was perceived as pure autocorrelation. It should be noted that the time dependence observed in the residuals in both cases is a first-order type of dependence. Both the Durbin–Watson statistic and the pattern of residuals indicate dependence between residuals in adjacent time periods. If the pattern of time dependence is other than first order the plot of residuals will still be informative. However, the Durbin—Watson statistic is not designed to measure higher-order time dependence and may not yield much valuable information.

As an example we consider the efforts of a company that produces and markets ski equipment in the United States to obtain a simple aggregate relationship of quarterly sales to some leading economic indicator. The indicator chosen is personal disposable income, PDI, in billions of current dollars. The initial model is

$$S_t = \beta_0 + \beta_1 (\text{PDI})_t + u_t,$$

where S_t is ski sales in period t in millions of dollars. Data for 10 years (40 quarters) is available and is listed in Table 6.6. The regression output is in Table 6.7 and a plot of residuals against time is given in Figure 6.5.

6.9.　EXAMINING RESIDUAL PLOTS

At first glance the results in Table 6.7 are encouraging. The proportion of variation in sales explained by PDI is .80. The marginal contribution of an additional dollar unit of PDI to sales is between \$166,503 and \$229,317 ($b_1 = .198$) with a confidence coefficient of 95%. In addition the Durbin–Watson statistic is 1.968 indicating no first-order autocorrelation.

It should be expected that PDI would explain a large proportion of the variation in sales since both variables are increasing over time. Therefore although the R^2 value of .80 is good, it should not be taken as a final evaluation of the model. Also, the Durbin–Watson value is in the acceptable range but it is clear from Figure 6.5 that there is some sort of time dependence of the residuals. We notice that residuals from the first and fourth quarters are positive while residuals from the second and third quarters are negative for all the years. Since skiing activities are affected by weather conditions we suspect that a seasonal effect has been overlooked. The pattern of residuals suggests that there are two seasons that have some bearing on ski sales: the second and third quarters which correspond to the warm weather season and the fourth and first quarters which correspond to the winter season when skiing is in full progress. This seasonal effect can be simply characterized by defining an indicator variable (dummy variable) that takes the value 1 for each winter quarter and is set equal to zero

Table 6.6.   Disposable income and ski sales for years 1964–1974.

	SALES	PDI
Q1/64	37.0	109
Q2/64	33.5	115
Q3/64	30.8	113
Q4/64	37.9	116
Q1/65	37.4	118
Q2/65	31.6	120
Q3/65	34.0	122
Q4/65	38.1	124
Q1/66	40.0	126
Q2/66	35.0	128
Q3/66	34.9	130
Q4/66	40.2	132
Q1/67	41.9	133
Q2/67	34.7	135
Q3/67	38.8	138
Q4/67	43.7	140
Q1/68	44.2	143
Q2/68	40.4	147
Q3/68	38.4	148
Q4/68	45.4	151
Q1/69	44.9	153
Q2/69	41.6	156
Q3/69	44.0	160
Q4/69	48.1	163
Q1/70	49.7	166
Q2/70	43.9	171
Q3/70	41.6	174
Q4/70	51.0	175
Q1/71	52.0	180
Q2/71	46.2	184
Q3/71	47.1	187
Q4/71	52.7	189
Q1/72	52.2	191
Q2/72	47.0	193
Q3/72	47.8	194
Q4/72	52.8	196
Q1/73	54.1	199
Q2/73	49.5	201
Q3/73	49.5	202
Q4/73	54.3	204

Table 6.7.   Ski sales against PDI

Variable	Coefficient	SE	t
PDI	0.198	0.0160	12.351
CONSTANT	12.392	2.539	4.880
$n = 40$	$R^2 = 0.8006$	$s = 3.019$	$d = 1.968$

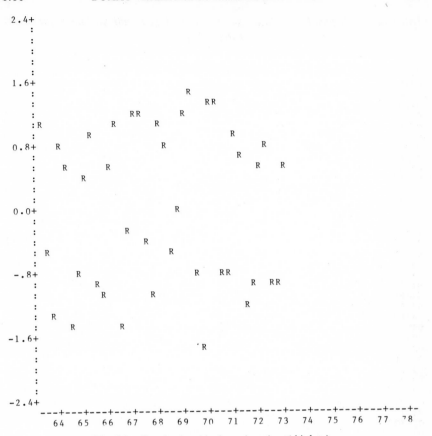

Fig. 6.5.   Graph of residuals against time (ski data).

for each summer quarter. (See Chapter 4.) The expanded data set is listed in Table 6.8.

6.10.   DUMMY VARIABLES TO REMOVE SEASONALITY

Using the additional seasonal variable, the model is expanded to be

$$S_t = \beta_0 + \beta_1 (\text{PDI})_t + \gamma Z_t + u_t,$$

where Z_t is the zero–one variable described above and γ is a parameter that measures the seasonal effect. Specifically the model represents the assumption that sales can be approximated by a linear function of PDI, one line for the winter season and one for the summer season. The lines

Table 6.8. *Disposable income, ski sales, and seasonal variables for years 1964–1974.*

	SALES	PDI	SEAS
Q1/64	37.0	109	1
Q2/64	33.5	115	0
Q3/64	30.8	113	0
Q4/64	37.9	116	1
Q1/65	37.4	118	1
Q2/65	31.6	120	0
Q3/65	34.0	122	0
Q4/65	38.1	124	1
Q1/66	40.0	126	1
Q2/66	35.0	128	0
Q3/66	34.9	130	0
Q4/66	40.2	132	1
Q1/67	41.9	133	1
Q2/67	34.7	135	0
Q3/67	38.8	138	0
Q4/67	43.7	140	1
Q1/68	44.2	143	1
Q2/68	40.4	147	0
Q3/68	38.4	148	0
Q4/68	45.4	151	1
Q1/69	44.9	153	1
Q2/69	41.6	156	0
Q3/69	44.0	160	0
Q4/69	48.1	163	1
Q1/70	49.7	166	1
Q2/70	43.9	171	0
Q3/70	41.6	174	0
Q4/70	51.0	175	1
Q1/71	52.0	180	1
Q2/71	46.2	184	0
Q3/71	47.1	187	0
Q4/71	52.7	189	1
Q1/72	52.2	191	1
Q2/72	47.0	193	0
Q3/72	47.8	194	0
Q4/72	52.8	196	1
Q1/73	54.1	199	1
Q2/73	49.5	201	0
Q3/73	49.5	202	0
Q4/73	54.3	204	1

are parallel; that is the marginal effect of changes in PDI is the same in both seasons. The level of sales as reflected by the intercept is different in each season (Figure 6.6).

Since Z_t is always 1 in the cold weather quarters and zero for warm quarters, we have

$$S_t = (\beta_0 + \gamma) + \beta_1 (\text{PDI})_t + u_t$$

for the winter season and for the summer season

$$S_t = \beta_0 + \beta_1 (\text{PDI})_t + u_t.$$

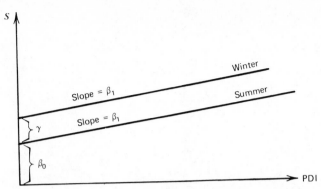

Fig. 6.6. Model for sales and PDI adjusted for season.

The regression results are summarized in Table 6.9 and the graph of residuals against time is found in Figure 6.7. We see that all indications of the seasonal pattern have been removed.

Furthermore, the precision of the estimated marginal effect of PDI increased. The confidence interval is now \$186,920 to \$210,440. Also the seasonal effect has been quantified and we can say that for a fixed level of PDI the winter season brings between \$4,759,484 and \$6,169,116 over the summer season (with confidence coefficient equal to 95%).

The ski data illustrates two important points concerning autocorrelation. First, the Durbin–Watson statistic is only sensitive to correlated errors when the correlation occurs between adjacent observations (first-order autocorrelation). In the ski data the first-order correlation is .02. The second-, fourth-, sixth-, and eighth-order correlations are $-.81$, $.76$, $-.71$, and $.73$, respectively. The Durbin–Watson test does not show significance in this case. There are other tests that may be used for the detection of higher-order autocorrelations, see Box and Pierce (1970). But in all cases, the graph of residuals will show the presence of time dependence in the error term when it exists.

Second, when autocorrelation is indicated the model should be refitted. Often the autocorrelation appears because a time-dependent variable is

Table 6.9. Ski sales against PDI and seasonal variables

Variable	Coefficient	SE	t
PDI	0.1987	0.00604	32.914
Seasonal	5.4643	0.0360	15.192
CONSTANT	9.5402	0.9748	9.786
$n=40$	$R^2=0.972$	$s=1.137$	$d=1.772$

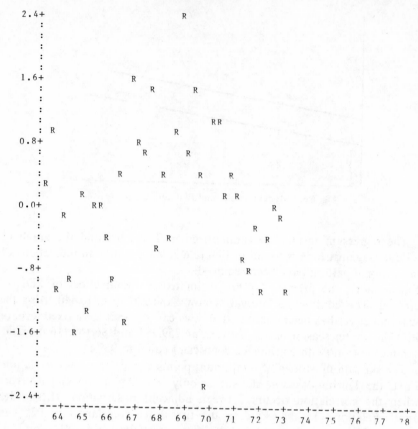

Fig. 6.7. Graph of residuals against time (ski data, with seasonal variables).

missing from the model. The inclusion of the omitted variable often removes the observed autocorrelation. Sometimes, however, no such variable is present. Then one has to make a differencing-type of transformation on the original variables to remove the autocorrelation.

REFERENCES

Box, G. E. P. and D. A. Pierce, Distribution of residual autocorrelation in autoregressive-integrated moving average time series models, *J. Amer. Stat. Assoc.*, **64**, 1509–1526 (1970).

Cochrane, D, and G. H. Orcutt, Application of least squares regression to relationships containing autocorrelated error terms, *J. Amer. Stat. Assoc.*, **44**, 32–61 (1949).

Johnston, J., *Econometric Methods*, McGraw-Hill, New York, 1972.

Kmenta, J., *Elements of Econometrics*, Macmillan, New York, 1971.

CHAPTER 7

Analysis of Collinear Data

7.1. INTRODUCTION

Interpretation of the multiple regression equation depends implicitly on the assumption that the explanatory variables are not strongly interrelated. It is usual to interpret a regression coefficient as measuring the change in the response variable when the corresponding explanatory variable is increased by one unit and all other explanatory variables are held constant. This interpretation may not be valid if there are strong linear relationships among the explanatory variables. It is always conceptually possible to increase the value of one variable in an estimated regression equation while holding the others constant. However, there may be no information about the result of such a manipulation in the estimation data. Moreover, it may be impossible to change one variable while holding all others constant in the process being studied. When these conditions exist, simple interpretation of the regression coefficient as a marginal effect is lost.

When there is a complete absence of linear relationship among the explanatory variables, they are said to be orthogonal. In most regression applications the explanatory variables are not orthogonal. Usually the lack of orthogonality is not serious enough to affect the analysis. However, in some situations the explanatory variables are so strongly interrelated that the regression results are ambiguous. Typically, it is impossible to estimate the unique effects of individual variables in the regression equation. The estimated values of the coefficients are very sensitive to slight changes in the data and to the addition or deletion of variables in the equation. The regression coefficients have large sampling errors which affect both inference and forecasting that is based on the regression model.

The condition of severe nonorthogonality is also referred to as the problem of collinear data, or multicollinearity. The problem can be extremely difficult to detect. It is not a specification error that may be uncovered by exploring regression residual. In fact, multicollinearity is not a modeling error. It is a condition of deficient data. In any event, it is important to know when multicollinearity is present and to be aware of its possible consequences. It is recommended that one should be extremely cautious about any and all substantive conclusions based on a regression analysis in the presence of multicollinearity.

This chapter focuses on three questions. First, how does multicollinearity affect statistical inference and forecasting? Second, how can multicollinearity be detected? And last, what can be done to resolve the difficulties associated with multicollinearity? When analyzing data, these questions cannot be answered separately. If multicollinearity is a potential problem, the three issues must be treated simultaneously by necessity.

The discussion begins with two examples. They have been chosen to demonstrate the effects of multicollinearity on inference and forecasting, respectively. A treatment of methods for detecting multicollinearity follows and the chapter concludes with a presentation of methods for resolving problems of multicollinearity. The obvious prescription to collect better data is considered, but the discussion is mostly directed at improving interpretation of the existing data. Alternatives to the ordinary least squares estimation method that perform efficiently in the presence of multicollinearity are considered in Chapter 8.

7.2. EFFECTS ON INFERENCE

This first example demonstrates the ambiguity that may result when attempting to identify important explanatory variables from among a linearly dependent collection of explanatory variables. The context of the example is borrowed from research on equal opportunity in public education as reported by Coleman and others. (Coleman, 1966; Mosteller, Moynihan, 1972).

In conjunction with the Civil Rights Act of 1964 the Congress of the United States ordered a survey "...concerning the lack of availability of equal educational opportunities for individuals by reason of race, color, religion or national origin in public educational institutions...". Data were collected from a cross section of school districts throughout the country. In addition to reporting summary statistics on variables such as level of student achievement and school facilities, regression analysis was used to try to establish factors that are the most important determinants of achievement. The data for this example consists of measurements taken in 1965 for 70 schools selected at random. The data consists of variables that measure student achievement, school facilities, and faculty credentials. The objective is to evaluate the effect of school inputs on achievement.

Assume that an acceptable index has been developed to measure those aspects of the school environment that would be expected to affect achievement. The index includes evaluations of the physical plant, teaching materials, special programs, training and motivation of the faculty, and so on. Achievement can be measured by using an index constructed from standardized test scores. There are also other variables that may affect the relationship between school inputs and achievement. Students' performances may be affected by their home environments and the influence of their peer group in the school. These variables must be accounted for in the analysis before the effect of school inputs can be evaluated. We assume that indexes have been constructed for these variables that are satisfactory for our purposes. The data is found in Table 7.1.

Adjustment for the two basic variables can be accomplished by using the regression model

$$ACHV = \beta_0 + \beta_1 \cdot FAM + \beta_2 \cdot PEER + \beta_3 \cdot SCHOOL + u. \qquad (7.1)$$

The contribution of the school variable can be tested using the t value for β_3. Recall that the t for β_3 tests whether SCHOOL is necessary in the equation when FAM and PEER are already included. Effectively, the model above is being compared to

$$ACHV = \beta_0 + \beta_1 \cdot FAM + \beta_2 \cdot PEER + u; \qquad (7.2)$$

that is, the contribution of the school variable is being evaluated after adjustment for FAM and PEER. Another view of the adjustment notion is obtained by manipulating Equation (7.1) to form

$$ACHV - \beta_1 \cdot FAM - \beta_2 \cdot PEER = \beta_0 + \beta_3 \cdot SCHOOL + u.$$

The left-hand side is an adjusted achievement index where adjustment is accomplished by subtracting the linear contributions of FAM and PEER. The equation is in the form of a regression of the adjusted achievement score on the SCHOOL variable. This representation is used only for the sake of interpretation. The estimated β's are obtained from the original model given in equation (7.1) above.

The regression results are summarized in Table 7.2 and a plot of the residuals against the predicted values of ACHV appears as Figure 7.1. Checking first the residual plot we see that there are no glaring indications of misspecification. The data observation located in the lower left of the graph has a residual value that is about 2.5 standard deviations from the mean of zero and should possibly be looked at more closely. However, when it is deleted from the sample, the regression results show almost no change. Therefore, the observation has been retained in the analysis.

Table 7.1. **Equal educational opportunity (EEO) data; standardized indexes**

ROW	FAM	PEER	SCHOOL	ACHV
* 1 *	0.60814	0.03509	0.16607	-0.43148
* 2 *	0.79369	0.47924	0.53356	0.79969
* 3 *	-0.82630	-0.61951	-0.78635	-0.92467
* 4 *	-1.25310	-1.21675	-1.04076	-2.19081
* 5 *	0.17399	-0.18517	0.14229	-2.84818
* 6 *	0.20246	0.12764	0.27311	-0.66233
* 7 *	0.24184	-0.09022	0.04967	2.63674
* 8 *	0.59421	0.21750	0.51876	2.35847
* 9 *	-0.61561	-0.48971	-0.63219	-0.91305
* 10 *	0.99391	0.62228	0.93368	0.59445
* 11 *	1.21721	1.00627	1.17381	1.21073
* 12 *	0.41436	0.71103	0.58978	1.87164
* 13 *	0.83782	0.74281	0.72154	-0.10178
* 14 *	-0.75512	-0.64411	-0.56986	-2.87949
* 15 *	-0.37407	-0.13787	-0.21770	3.92590
* 16 *	1.40353	1.14085	1.37147	4.35084
* 17 *	1.64194	1.29229	1.40269	1.57922
* 18 *	-0.31304	-0.07980	-0.21455	3.95689
* 19 *	1.28525	1.22441	1.20428	1.09275
* 20 *	-1.51938	-1.27565	-1.36598	-0.62389
* 21 *	-0.38224	-0.05353	-0.35560	-0.63654
* 22 *	-0.19186	-0.42605	-0.53718	-2.02659
* 23 *	1.27649	0.81427	0.91967	-1.46692
* 24 *	0.52310	0.30720	0.47231	3.15078
* 25 *	-1.59810	-1.01572	-1.48315	-2.18938
* 26 *	0.77914	0.87771	0.76496	1.91715
* 27 *	-1.04745	-0.77536	-0.91397	-2.71428
* 28 *	-1.63217	-1.47709	-1.71347	-6.59852
* 29 *	0.44328	0.60956	0.32833	0.65101
* 30 *	-0.24972	0.07876	-0.17216	-0.13772
* 31 *	-0.33480	-0.39314	-0.37198	-2.43959
* 32 *	-0.20680	-0.13936	0.05626	-3.27802
* 33 *	-1.99375	-1.69587	-1.87838	-2.48058
* 34 *	0.66475	0.79670	0.69865	1.88639
* 35 *	-0.27977	0.10817	-0.26450	5.06459
* 36 *	-0.43990	-0.66022	-0.58490	1.96335
* 37 *	-0.05334	-0.02396	-0.16795	0.26274
* 38 *	-2.06699	-1.31832	-1.72082	-2.94593
* 39 *	-1.02560	-1.15858	-1.19420	-1.38628
* 40 *	0.45847	0.21555	0.31347	-0.20797
* 41 *	0.93979	0.63454	0.69907	-1.07820
* 42 *	-0.93238	-0.95216	-1.02725	-1.66386
* 43 *	-0.35988	-0.30693	-0.46232	0.58117

Source: The data is not authentic. It has been generated by the authors for the purpose of this example.

From Table 7.2 we see that about 20% of the variation in achievement score is explained by the three regressors jointly ($R^2 = .206$). The F value is 5.72 based on 3 and 66 degrees of freedom and is significant at better than the .01 level. Therefore even though the total explained variation is estimated at only 20%, it is accepted that the regressors are valid explanatory variables. However, the individual t values are all small. In total, the

Table 7.1. (continued)

ROW	FAM	PEER	SCHOOL	ACHV
* 44 *	-0.00518	0.35985	0.02485	1.37447
* 45 *	-0.18892	-0.07959	0.01704	-2.82687
* 46 *	0.87271	0.47644	0.57036	3.86363
* 47 *	-2.06993	-1.82915	-2.16738	-2.64141
* 48 *	0.32143	-0.25961	0.21632	0.05387
* 49 *	-1.42382	-0.77620	-1.07473	0.50763
* 50 *	-0.07852	-0.21347	-0.11750	0.64347
* 51 *	-0.14925	-0.03192	-0.36598	2.49414
* 52 *	0.52666	0.79149	0.71369	0.61955
* 53 *	-1.49102	-1.02073	-1.38103	0.61745
* 54 *	-0.94757	-1.28991	-1.24799	-1.00743
* 55 *	0.24550	0.83794	0.59596	-0.37469
* 56 *	-0.41630	-0.60312	-0.34951	-2.52824
* 57 *	1.38143	1.54542	1.59429	0.02372
* 58 *	1.03806	0.91637	0.97602	2.51077
* 59 *	-0.88639	-0.47652	-0.77693	-4.22716
* 60 *	1.08655	0.65700	0.89401	1.96847
* 61 *	-1.95142	-1.94199	-1.89645	1.25668
* 62 *	2.83384	2.47398	2.79222	-0.16848
* 63 *	1.86753	1.55229	1.80057	-0.34158
* 64 *	-1.11172	-0.69732	-0.80197	-2.23973
* 65 *	1.41958	1.11481	1.24558	3.62654
* 66 *	0.53940	0.16182	0.33477	0.97034
* 67 *	0.22491	0.74800	0.66182	3.16093
* 68 *	1.48244	1.47079	1.54283	-1.90801
* 69 *	2.05425	1.80369	1.90066	0.64598
* 70 *	1.24058	0.64484	0.87372	-1.75915

summary statistics say that the three regressors taken together are important, but from the t values it follows that any one regressor may be deleted from the model provided the other two are retained.

These results are typical of a situation where extreme multicollinearity is present. The explanatory variables are so highly correlated that each one may serve as a proxy for the others in the regression equation without affecting the total explanatory power. The low t values confirm that any one of the explanatory variables may be dropped from the equation. Hence, the regression analysis has failed to provide any information for evaluating the importance of school inputs on achievement. The culprit is clearly multicollinearity. The simple correlations between FAM and

Table 7.2. Regression results for EEO data

Variable	Coefficient	SE	t
FAM	1.101	1.411	.780
PEER	2.322	1.481	1.568
SCHOOL	-2.281	2.220	-1.027
CONSTANT	-.070	.251	.279
$n = 70$	$R^2 = .206$	$s = 2.070$	

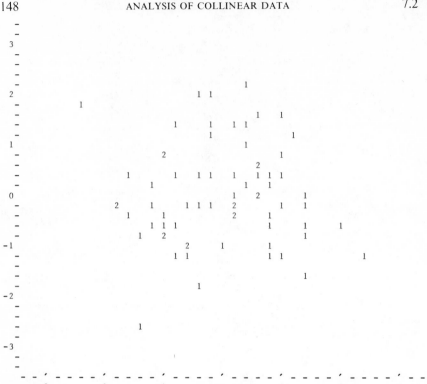

Fig. 7.1. Standardized residuals against fitted values of ACHV.

PEER, FAM and SCHOOL, and PEER and SCHOOL are .960, .986, and .982, respectively, showing strong linear relationships among all pairs of explanatory variables.

The multicollinearity in this instance could have been expected. It is the nature of these three variables that each one is determined by and helps to determine the others. It is not unreasonable to conclude that there are not three variables but in fact only one.

Unfortunately, that conclusion does not help to answer the original question about the effects of school facilities on achievement. There remain two possibilities. First, multicollinearity may be present because the sample data is deficient, but can be improved with additional observations. Second, multicollinearity may be present because the interrelationships among the variables are an inherent characteristic of the process under investigation. Both situations are discussed in the following paragraphs.

In the first case the sample should have been selected to insure that the correlations between the explanatory variables were not large. For example, consider FAM and SCHOOL. A scatter plot of FAM and SCHOOL for the sample is given in Figure 7.2. All the observations lie close to the straight line through the average values of FAM and SCHOOL. There are no schools in the sample with values in the upper left or lower right of the graph. Hence there is no information in the sample on achievement when the value of FAM is high and SCHOOL is low, or FAM is low and SCHOOL is high. But it is only with data collected under these two conditions that the individual effects of FAM and SCHOOL on ACHV can be determined. For example, assume that there were some observations in the upper left quadrant of the graph. Then it would at least be possible to compare average ACHV for low and high values of SCHOOL when FAM is held constant.

Adding the third explanatory value, PEER, to the model means that

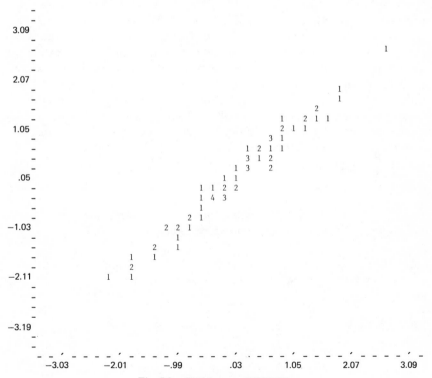

Fig. 7.2. FAM versus SCHOOL.

there are eight distinct combinations of data that should be included in the sample. Using + to represent a value above the average and − to represent a value below the average, the eight possibilities are represented in Table 7.3.

The large correlations that were found in the analysis suggest that only combinations 1 and 8 are represented in the data. If the sample turned out this way by chance, the prescription for resolving the multicollinearity problem is to collect additional data on some of the other combinations. For example, data based on combinations 1 and 2 alone could be used to evaluate the effect of SCHOOL on ACHV holding FAM and PEER at a constant level, both above average. If these were the only combinations represented in the data, the analysis would consist of the simple regression of ACHV against SCHOOL. The results would give only a partial answer, namely an evaluation of the school–achievement relationship when FAM and PEER are both above average.

The prescription for additional data as a way to resolve multicollinearity is not a panacea. It is often not possible to collect more data because of constraints on budgets, time, and staff. It is always better to be aware of impending data deficiencies beforehand. Whenever possible, the data should be collected according to design. Unfortunately, prior design is not always feasible. In surveys, or observational studies such as the one being discussed, the values of the explanatory variables are usually not known until the sampling unit is selected for the sample and some costly and time-consuming measurements are developed. Following this procedure, it is fairly difficult to insure that a balanced sample will be obtained.

The second reason that multicollinearity may appear is because the relationships among the variables are an inherent characteristic of the process being sampled. If FAM, PEER, and SCHOOL exist in the popula-

Table 7.3. Data combinations for three explanatory variables

	Variable		
Combination	FAM	PEER	SCHOOL
1	+	+	+
2	+	+	−
3	+	−	+
4	−	+	+
5	+	−	−
6	−	+	−
7	−	−	+
8	−	−	−

tion only as data combinations 1 and 8 of Table 7.3, then it is impossible to estimate the individual effects of these variables on achievement. The only recourse for continued analysis of these effects would be to search for underlying causes that may explain the interrelationships of the explanatory variables. Through this process, one may discover other variables that are more basic determinants affecting equal opportunity in education and achievement.

7.3. EFFECTS ON FORECASTING

We shall examine the effects of multicollinearity in forecasting, when the forecasts are based on a multiple regression equation. A historical data set with observations indexed by time is used to estimate the regression coefficients. Forecasts of the response variable are produced by using future values of the explanatory variables in the estimated regression equation. The future values of the explanatory variables must be known or forecasted from other data and models. We shall not treat the uncertainty in the forecasted explanatory variables. In our discussion, it is assumed that the future values of the explanatory variables are given.

We have chosen an example based on aggregate data concerning import activity in the French economy. The data has been analyzed by Malinvaud (1966). Our discussion follows his presentation. The variables are imports (IMPORT), domestic production (DOPROD), stock formation (STOCK), and domestic consumption (CONSUM), all measured in milliards of French francs for the years 1949 through 1966. The data is given in Table 7.4. The model being considered is

$$IMPORT = \beta_0 + \beta_1 \cdot DOPROD + \beta_2 \cdot STOCK + \beta_3 \cdot CONSUM + u. \quad (7.3)$$

The regression results appear as Table 7.5. The plot of residuals against time, Figure 7.3, shows a distinctive pattern suggesting that the model is not well specified. Even though multicollinearity appears to be present ($R^2 = 0.973$, and all t values small), it should not be pursued further in this model. Multicollinearity should only be attacked after the model specification is satisfactory. The difficulty with the model is that the European Common Market began operations in 1960 causing changes in import-export relationships. Since our objective in this chapter is to study the effects of multicollinearity, we shall not complicate the model by attempting to capture the behavior after 1959. We shall assume that it is now 1960 and look only at the 11 years, 1949-1959. The regression results for that data are summarized in Table 7.6. The residual plot is now satisfactory (Figure 7.4).

Table 7.4.   *Data on French economy;* IMPORT *data (milliards of French francs)*

YEAR	IMPORT	DOPROD	STOCK	CONSUM
49	15.9	149.3	4.2	108.1
50	16.4	161.2	4.1	114.8
51	19.0	171.5	3.1	123.2
52	19.1	175.5	3.1	126.9
53	18.8	180.8	1.1	132.1
54	20.4	190.7	2.2	137.7
55	22.7	202.1	2.1	146.0
56	26.5	212.4	5.6	154.1
57	28.1	226.1	5.0	162.3
58	27.6	231.9	5.1	164.3
59	26.3	239.0	0.7	167.6
60	31.1	258.0	5.6	176.8
61	33.3	269.8	3.9	186.6
62	37.0	288.4	3.1	199.7
63	43.3	304.5	4.6	213.9
64	49.0	323.4	7.0	223.8
65	50.3	336.8	1.2	232.0
66	56.6	353.9	4.5	242.9

Source: *Statistical Methods in Econometrics* by E. Malinvaud, Rand McNally, Chicago, 1968.

The proportion of explained variation is high, .99. However, the coefficient of DOPROD is negative and not statistically significant which is contrary to prior expectation. We believe that if STOCK and CONSUM were held constant, an increase in DOPROD would cause an increase in IMPORT probably for raw materials or manufacturing equipment. Multicollinearity is a possibility here and in fact is the case. The simple correlation between CONSUM and DOPROD is .997. Upon further investigation it turns out that CONSUM has been about 2/3 of DOPROD throughout the 11-year period. The estimated relationship between the two quantities is

$$CONSUM = 6.258 + .686 \cdot DOPROD.$$

Even in the presence of such severe multicollinearity the regression

Table 7.5.   *Import data: 1949-1966*

Variable	Coefficient	SE	t
DOPROD	.032	.187	.171
STOCK	.414	.322	1.286
CONSUM	.243	.285	.853
CONSTANT	− 19.730	4.125	− 4.783
$n = 18$	$R^2 = .973$	$s = 2.258$	
		$F(3, 14) = 168.45$	

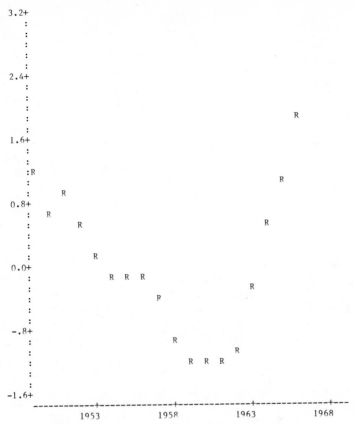

Fig. 7.3.   Standarized residuals versus Time; 1949–1966.

Table 7.6.   Regression results, Import data: 1949-1959

				Correlation matrix			
Variable	coefficient	SE	t	DOPROD	STOCK	CONSUM	IMPORT
DOPROD	− .051	.070	− .728	1	.026	.997	.965
STOCK	.587	.095	6.179		1	.036	.251
CONSUM	.287	.102	2.813			1	.972
CONSTANT	− 10.130	1.212	− 8.358				−
$n=11$	$R^2 = .992$	$s = .489$					

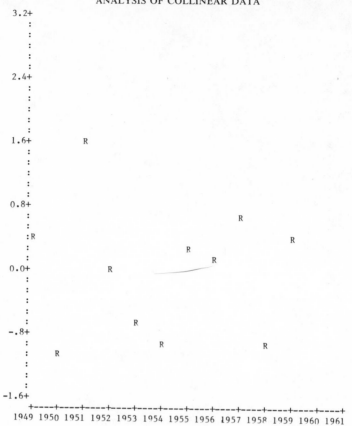

Fig. 7.4. Standardized residuals versus time; 1949–1959.

equation may produce some good forecasts. The forecasting equation is

$$IMPORT = -10.13 - .051 \cdot DOPROD + .587 \cdot STOCK + .287 \cdot CONSUM.$$

Recall that the fit to the historical data is very good and the residual variation appears to be purely random. In order to forecast we must be confident that the character and strength of the overall relationship will hold into future periods. This matter of confidence is a problem in all forecasting models whether or not multicollinearity is present. For the purpose of this example we shall assume that the overall relationship does hold into future periods.* Implicit in this assumption is the relationship

---

* For the purpose of convenient exposition we ignore the difficulties that arise because of our previous finding that the formation of the Common Market has altered the relationship since 1960. But we are impelled to advise the reader that changes in structure make forecasting a very delicate endeavor even when the historical fit is excellent.

between DOPROD and CONSUM. The forecast will be accurate as long as the future values of DOPROD, STOCK, and CONSUM are given with CONSUM $\doteq 2/3$ DOPROD.

For example, let us forecast the change in IMPORT next year corresponding to an increase in DOPROD of 10 units while holding STOCK and CONSUM at their current levels. The resulting forecast is

$$IMPORT_{1960} = IMPORT_{1959} - .51$$

or a decrease in IMPORT. However, if the relationship between DOPROD and CONSUM is kept intact, CONSUM will increase by 15 units and the forecasted result is

$$IMPORT_{1960} = IMPORT_{1959} - .51 + 4.305 = IMPORT_{1959} + 3.795.$$

IMPORT actually increases by 3.795 units, a more satisfying result and probably a better forecast. The case where DOPROD increases alone corresponds to a change in the basic structure of the data that was used to estimate the model parameters and cannot be expected to produce meaningful forecasts.

In summary, the two examples demonstrate that nonorthogonal data can seriously limit the use of regression analysis for inference and forecasting. Extreme care is required when attempting to interpret regression results when multicollinearity is suspected. In the next section we discuss methods for detecting extreme collinearity among explanatory variables.

7.4. DETECTION OF MULTICOLLINEARITY

In the preceding examples some of the ideas for detecting multicollinearity were already introduced. In this section, we review those ideas and introduce additional criteria which indicate collinearity. Multicollinearity is associated with unstable estimated regression coefficients. This situation results from the presence of strong linear relationships among the explanatory variables. It is not a problem of misspecification. Therefore, the empirical investigation of problems that result from a collinear data set should begin only after the model has been satisfactorily specified. However, there may be some indications of multicollinearity that are encountered during the process of adding, deleting, and transforming variables or data points in search of the good model. Indication of multicollinearity that appear as instability in the estimated coefficients are as follows:

1. large changes in the estimated coefficients when a variable is added or deleted,

2. large changes in the coefficients when a data point is altered or dropped.

Once the residual plots indicate that the model has been satisfactorily specified, multicollinearity may be present if

3. the algebraic signs of the estimated coefficients do not conform to prior expectations or

4. coefficients of variables that are expected to be important have large standard errors.

For the IMPORT data discussed previously, the coefficient of DOPROD was negative and not significant. Both results are contrary to prior expectations. The effects of dropping or adding a variable can be seen in Table 7.7. There we see that the presence or absence of certain variables has a large effect on the other coefficients. For the EEO data (Table 7.1) the algebraic signs are all correct, but their standard errors are so large that none of the coefficients are statistically significant. It was expected that they would all be important.

The presence of multicollinearity is also indicated by the size of the correlation coefficients that exist among the explanatory variables. A large correlation between a pair of explanatory variables indicates a strong linear relationship between those two variables. The correlations for the EEO data (p.148) are large for all pairs of explanatory variables. For the IMPORT data, the correlation coefficient between DOPROD and CONSUM is .997.

The source of multicollinearity may be more subtle than a simple relationship between two variables. A linear relation can involve many of the explanatory variables. It may not be possible to detect such a relationship with a simple correlation coefficient. As an example, we shall look at an analysis of the effects of advertising expenditures (A), promotion expenditures (P), and sales expense (SE) on the aggregate sales of a firm.

Table 7.7 Regression coefficients for all possible regressions IMPORT *data*

Regression	Variable		
	DOPROD	STOCK	CONSUM
1	.146	–	–
2	–	.691	–
3	–	–	.214
4	.145	.622	–
5	– .109	–	.372
6	–	.596	.212
7	– .051	.587	.287

The data (Table 7.8) represents a period of 23 years during which the firm was operating under fairly stable conditions. The proposed regression model is

$$S_t = \beta_0 + \beta_1 A_t + \beta_2 P_t + \beta_3 SE_t + \beta_4 A_{t-1} + \beta_5 P_{t-1} + u_t. \tag{7.4}$$

The regression results are found in Table 7.9. The plot of residuals versus fitted values of sales and time in Figures 7.5 and 7.6, as well as other plots of the residuals versus the independent variables (not shown), do not suggest any problems of misspecification. Furthermore, the correlation coefficients between the explanatory variables are small (Table 7.9). However, if we do a little experimentation to check the stability of the coefficients by dropping the contemporaneous advertising variable from the model, many things change. The coefficient of P_t drops from 8.37 to 3.83; the coefficients of lagged advertising and lagged promotions change signs and are no longer statistically significant. But the coefficient of sales expense is stable and R^2 does not change much. The evidence suggests that there is some type of relationship involving the contemporaneous and lagged values of A and P. The regression of A_t on P_t, A_{t-1}, and P_{t-1} returns an R^2 of .973. The equation takes the form

$$A_t = 4.63 - .87 P_t - .86 A_{t-1} - .95 P_{t-1}.$$

Upon further investigation into the operations of the firm, it was discovered that close control was exercised over the expense budget during those 23 years of stability. In particular, there was an approximate rule imposed on the budget that the sum of A_t, A_{t-1}, P_t, and P_{t-1} was to be held to approximately 5 units over every 2-year period. The relationship

$$A_t + P_t + A_{t-1} + P_{t-1} = 5$$

is the cause of the multicollinearity. We see that a complete search for multicollinearity includes checking the values of R^2 in the multiple regressions of each explanatory variable against all other explanatory variables.

7.5.   PRINCIPAL COMPONENTS IN DETECTION OF MULTICOLLINEARITY

The indicators of multicollinearity that have been described can all be obtained using standard regression computations. There is another, more unified way to analyze multicollinearity which requires some calculations that are not usually included in standard regression packages. The analysis follows from the fact that every linear regression model can be restated in terms of a set of orthogonal explanatory variables. These new variables are

Table 7.8.   *Annual data on advertising, promotions, sales expenses, and sales (in millions of dollars)*

ROW	A	P	SE	A-1	P-1	S
* 1 *	1.98786	1.0	0.30	2.01722	0.0	20.11371
* 2 *	1.94418	0.0	0.30	1.98786	1.0	15.10439
* 3 *	2.19954	0.8	0.35	1.94418	0.0	18.68375
* 4 *	2.00107	0.0	0.35	2.19954	0.8	16.05173
* 5 *	1.69292	1.3	0.30	2.00107	0.0	21.30101
* 6 *	1.74334	0.3	0.32	1.69292	1.3	17.85004
* 7 *	2.06907	1.0	0.31	1.74334	0.3	18.87558
* 8 *	1.01709	1.0	0.41	2.06907	1.0	21.26599
* 9 *	2.01906	0.9	0.45	1.01709	1.0	20.48473
* 10 *	1.06139	1.5	0.45	2.01906	1.0	20.54032
* 11 *	1.45999	1.5	0.50	1.06139	0.9	26.18441
* 12 *	1.87511	0.8	0.65	1.45999	1.5	21.71606
* 13 *	2.27109	0.0	0.65	1.87511	0.8	28.69595
* 14 *	1.11191	1.2	0.65	2.27109	0.0	25.83720
* 15 *	1.77407	1.0	0.65	1.11191	1.2	29.31987
* 16 *	0.95878	1.0	0.62	1.77407	1.0	24.19041
* 17 *	1.98930	0.7	0.60	0.95878	1.0	26.58966
* 18 *	1.97111	0.1	0.60	1.98930	1.0	22.24466
* 19 *	2.26603	1.0	0.61	1.97111	0.7	24.79944
* 20 *	1.98346	0.7	0.60	2.26603	0.7	21.19105
* 21 *	2.10054	0.1	0.58	1.98346	0.1	26.03441
* 22 *	1.06815	1.0		2.10054	1.0	27.39304

158

Table 7.9. Regression results, sales, advertising, promotion data

Variable	Coefficient	SE[a]	t	A	P	SE	A_{-1}	P_{-1}	S
						Correlation matrix			
A	5.3606	4.0277	1.331	1	−.35695	−.12852	−.13974	−.49599	−.17041
P	8.3722	3.5864	2.334		1	.06259	−.31646	−.29636	.54018
SE	22.5210	2.1424	10.512			1	−.16643	.20811	.81087
A_{-1}	3.8545	3.5778	1.077				1	−.35776	−.30516
P_{-1}	4.1247	3.8952	1.059					1	−.05204
CONSTANT	−14.1900	18.7150	−.758						−

$n = 22$ $R^2 = .9169$ $s = 1.3200$

[a]Standard error.

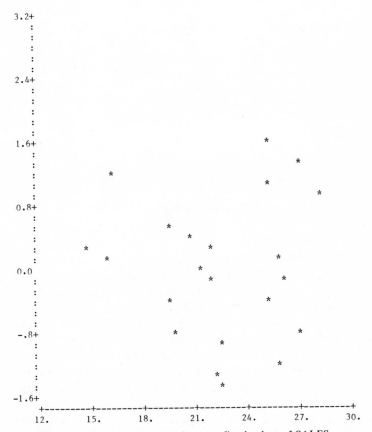

Fig. 7.5. Standardized residuals versus fitted values of SALES.

159

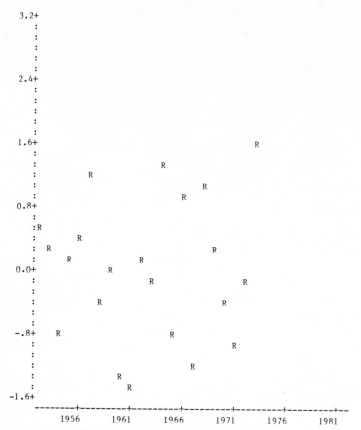

```
3.2+
   :
   :
   :
   :
2.4+
   :
   :
   :
   :
   :
1.6+                                              R
   :
   :                        R
   :          R
   :                              R
   :                        R
0.8+
   :
   :R
   :          R
   : R                          R
   :  R    R        R
0.0+          R
   :                R                R
   :
   :          R              R
   :
-.8+   R              R
   :                            R
   :
   :                  R
   :        R
   :          R
-1.6+
     --------+---------+---------+---------+---------+---------+--
         1956      1961      1966      1971      1976      1981
```

Fig. 7.6.   Standardized residuals versus time.

obtained as linear combinations of the original explanatory variables. They are referred to as the principal components (Kendall, 1957; Press, 1972) of the explanatory set of variables. The principal components lack simple interpretation since each is, in a sense, a mixture of the original variables. However, these new variables provide a unified approach for obtaining information about multicollinearity and serve as the basis of one of the alternative estimation techniques described in the next chapter.

In order to develop the method of principal components, first note that the basis of any analysis of multicollinearity is found in the structure of the correlations among the explanatory variables. Since correlations are unaffected by shifting or scaling the data, it is both sufficient and convenient to deal with standarized variables. A standarized variable is obtained by subtracting the average value from each observation and dividing by the standard deviation of the observations. The average value of a standardized variable is zero. The variance is equal to unity and the covariance between standardized variables are the correlations of the unstandardized data. For notational convenience, the standardized variable names will be in lower case letter.

Recall that a set of variables is said to be orthogonal if there exists no linear relationships among the variables. If the standardized explanatory variables are orthogonal, their matrix of variances and covariances consists of one for the diagonal elements and zero for the off-diagonal elements.* For example, if doprod, stock, and consum of the IMPORT data were orthogonal, the covariance matrix would be of the form

	doprod	stock	consum	
doprod	1	0	0	
stock	0	1	0	.
consum	0	0	1	

The actual covariance matrix of the standardized explanatory variables from the IMPORT data is

	doprod	stock	consum	
doprod	1	.026	.997	
stock	.026	1	.036	.
consum	.997	.036	1	

The standardized variables of the IMPORT data are transformed** to the principal components, named Z_1, Z_2, and Z_3 by the equations

$$
\begin{aligned}
Z_1 &= .7063 \cdot \text{doprod} &+ .0435 \cdot \text{stock} &+ .7065 \cdot \text{consum}, \\
Z_2 &= -.0357 \cdot \text{doprod} &+ .9990 \cdot \text{stock} &- .0258 \cdot \text{consum}, \quad (7.5) \\
Z_3 &= -.7070 \cdot \text{doprod} &- .0070 \cdot \text{stock} &+ .7072 \cdot \text{consum}.
\end{aligned}
$$

---

*The variances and covariances of a set of p variables can be neatly displayed as a square array of numbers called a matrix. The element found at the intersection of the ith row and jth column is the covariance of variable i with variable j. The diagonal element of the ith row is the variance of variable i.

**A description of this technique employing matrix algebra is given in the Appendix to this chapter.

The covariance matrix of the new variables* is of the form

$$
\begin{array}{ccc}
 & Z_1 \quad Z_2 \quad Z_3 \\
\begin{array}{c} Z_1 \\ Z_2 \\ Z_3 \end{array}
\begin{bmatrix}
\lambda_1 & 0 & 0 \\
0 & \lambda_2 & 0 \\
0 & 0 & \lambda_3
\end{bmatrix}
\end{array}
$$

with $\lambda_1 \geqslant \lambda_2 \geqslant \lambda_3$. The λ's are the variances of the new variables, Z_1, Z_2, Z_3. If the λ's are all equal to unity, the original variables are orthogonal. If any one of the λ's is exactly equal to zero, there is a perfect linear relationship among the original variables which is the extreme case of multicollinearity. If one of the λ's is much smaller than the others (and near zero), multicollinearity is present. (The λ's are also known as the characteristic roots of the correlation matrix of the original explanatory variables.)

For the IMPORT data, $\lambda_1 = 1.999$, $\lambda_2 = .998$, and $\lambda_3 = .003$. The small characteristic root points to multicollinearity. The other data sets considered in this chapter also have informative roots. For the EEO data, $\lambda_1 = 2.952$, $\lambda_2 = .040$, and $\lambda_3 = .008$. For the advertising data, $\lambda_1 = 1.701$, $\lambda_2 = 1.288$, $\lambda_3 = 1.145$, $\lambda_4 = .859$, and $\lambda_5 = .007$. In each case, the presence of a small characteristic root is indicative of multicollinearity.

One additional piece of information is available through this type of analysis. Since the λ's are variances of the principal components, if λ is approximately zero, the corresponding principal component is approximately equal to a constant. It follows that the equation defining the principal component gives some idea about the type of relationship among the explanatory variables that is causing multicollinearity. For example, in the IMPORT data, $\lambda_3 = .003 \doteq 0$. Therefore, Z_3 is approximately constant. The constant is the mean value of Z_3 which is zero. The principal components all have means of zero since they are linear functions of the standardized variables and each standardized variable has a zero mean. Therefore,

$$
Z_3 = -.7070 \cdot \text{doprod} - .0070 \cdot \text{stock} + .7072 \cdot \text{consum} \doteq 0.
$$

Rearranging the terms yields

$$
\text{consum} \doteq \text{doprod}, \tag{7.6}
$$

---

*The variances are the characteristic roots (eigenvalues, latent roots) of the correlation matrix of the original variables. The constants defining the transformation to principal components are the components of the characteristic vectors. See the Appendix.

where $.7070/.7072$ has been taken to be equal to unity and $.007/.7072$ has been approximated as zero. Equation (7.6) represents the approximate relationship that exists between the standardized versions of CONSUM and DOPROD. This result is consistent with our previous finding based on the high simple correlation between CONSUM and DOPROD, $r = .997$. Since λ_3 is the only small characteristic root, the analysis of principal components tells us that the dependence structure among the explanatory variables as reflected in the data is no more complex than the simple relationship between consum and doprod as given in Equation (7.6).

For the advertising data (ADVER), the smallest characteristic root is $\lambda_5 = .007$. The corresponding principal component is

$$Z_5 = .514a + .489p - .010se + .428a_{-1} + .559p_{-1}. \tag{7.7}$$

Setting Z_5 to zero and solving for a leads to the approximate relationship,

$$a \doteq -.951p - .833a_{-1} - 1.087p_{-1}. \tag{7.8}$$

This equation reflects our earlier findings about the relationship between A, P, A_{-1}, and P_{-1}. Furthermore, since $\lambda_4 = .859$ and the other λ values are all larger, we can be confident that the relationship involving A, A_{-1}, P, and P_{-1} in Equation (7.8) is the only source of multicollinearity in the data.

Throughout this section, investigations concerning the presence of multi-collinearity have been based on judging the magnitudes of various indicators, either a correlation coefficient or a characteristic root. Although we speak in terms of large and small, there is no way to determine threshold values that will precisely define those words. Size is relative and can be used to give an indication either that everything seems to be in order or that something is amiss. The only reasonable criterion for judging size is to decide whether the ambiguity resulting from the perceived multicollinearity is of material importance in the underlying problem.

7.6.   CORRECTION FOR MULTICOLLINEARITY: IMPOSING CONSTRAINTS

We have noted that multicollinearity is a condition associated with deficient data and not due to misspecification of the model. It is assumed that the form of the model has been carefully structured and that the residuals are acceptable before questions of multicollinearity are considered. Since it is usually not practical and often impossible to improve the data, we shall focus our attention on methods of better interpretation of the given data than would be available from a direct application of least squares. In this section, rather than trying to interpret individual regression coefficients, we shall attempt to identify and estimate informative linear functions of the regression coefficients. Alternative estimating methods for the individual coefficients are treated in the next chapter.

Before turning to the problem of searching the data for informative linear functions of the regression coefficients, one additional point concerning model specification must be discussed. A subtle step in specifying a relationship that can have a bearing on multicollinearity is acknowledging the presence of theoretical relationships among the regression coefficients. For example in the model for the IMPORT data,

$$IMPORT = \beta_0 + \beta_1 \cdot DOPROD + \beta_2 \cdot STOCK + \beta_3 \cdot CONSUM + u,$$

one may argue that the marginal effects of DOPROD and CONSUM are equal. That is, on the basis of economic reasoning, and before looking at the data, it is decided the $\beta_1 = \beta_3$ or equivalently, $\beta_1 - \beta_3 = 0$. Following the method described in Chapter 3, the common value of β_1 and β_3 is estimated by regressing IMPORT on STOCK and a new variable constructed as NEWVAR = DOPROD + CONSUM. The new variable has significance only as a technical manipulation to extract an estimate of the common value of β_1 and β_3. The results of the regression appear in Table 7.10. The correlation between the two explanatory variables, STOCK and NEWVAR, is .0299 and the characteristic roots are $\lambda_1 = 1.030, \lambda_2 = .970$. There is no longer any indication of multicollinearity. The residual plots against time and the fitted values indicate that there are no other problems of specification (Figures 7.7 and 7.8, respectively). The estimated model is

$$IMPORT = -9.007 + .086 \cdot DOPROD + .612 \cdot STOCK + .086 \cdot CONSUM.$$

Note that following the methods outlined in Chapter 3, it is also possible to test the constraint, $\beta_1 = \beta_3$, as a hypothesis. Even though the argument for $\beta_1 = \beta_3$ may have been imposed on the basis of existing theory, it is still interesting to evaluate the effect of the constraint on the explanatory

Table 7.10.   Regression results of IMPORT *data with constraint;* $\beta_1 = \beta_3$

Variable	Coefficient	SE	t
STOCK	.612	.109	5.60
NEWVAR	.086	.003	24.23
CONSTANT	--9.007		
	$R^2 = .987$	$s = .569$	

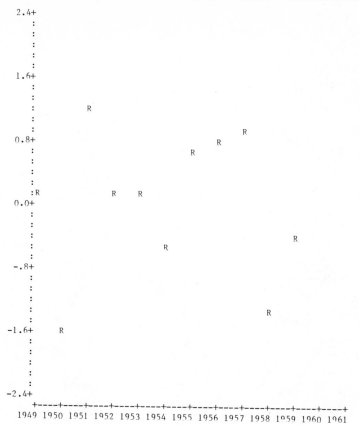

Fig. 7.7.   Standardized residuals versus time; IMPORT data: CONSTRAINT, $\beta_1 = \beta_3$.

power of the full model. The values of R^2 for the full and restricted models are .992 and .987, respectively. The F ratio for testing $H_0(\beta_1 = \beta_3)$ is 3.36 with 1 and 7 degrees of freedom. Both results suggest that the constraint is consistent with the data.

The constraint that $\beta_1 = \beta_3$ is, of course, only one example of the many types of constraints that may be used when specifying a regression model. The general class of possibilities is found in the set of linear constraints described in Chapter 3. Constraints are usually justified on the basis of underlying theory. They may often resolve what appears to be a problem of multicollinearity. In addition, any particular constraint may be viewed as a testable hypothesis and judged by the methods described in Chapter 3.

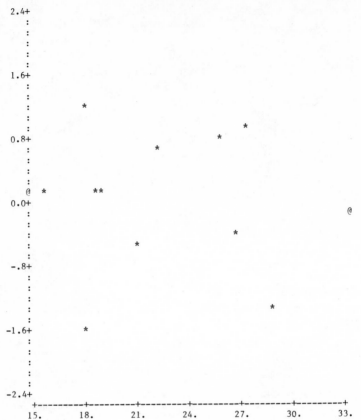

```
2.4+
   :
   :
   :
   :
   :
1.6+
   :
   :                    *
   :
   :                                              *
   :                                        *
0.8+                              *
   :                       *
   :
   :
   :
   @     *        **
0.0+                                                                              @
   :
   :
   :                                    *
   :                          *
   :
-.8+
   :
   :
   :
   :                                              *
   :
-1.6+          *
   :
   :
   :
   :
-2.4+
    +---------+---------+---------+---------+---------+---------+
    15.       18.       21.       24.       27.       30.       33.
```

Fig. 7.8. Standardized residuals against fitted values of IMPORT: CONSTRAINT, $\beta_1 = \beta_3$.

7.7. SEARCHING FOR LINEAR FUNCTIONS OF THE β's

We assume that the model

$$y = \beta_0 + \beta_1 x_1 + \cdots + \beta_p x_p + u$$

has been carefully specified so that the regression coefficients appearing there are of primary interest for policy analysis and decision making. We have seen that the presence of multicollinearity may prevent individual β's from being accurately estimated. However, as demonstrated below, it is always possible to estimate some linear functions of the β's accurately (Silvey, 1969). The obvious questions are: Which linear functions can be estimated, and of those that can be estimated, which are of interest in the analysis? In this section we shall use the data to help identify those linear functions that can be accurately estimated and, at the same time, have some value in the analysis.

First we shall demonstrate in an indirect way that there are always linear functions of the β's that can be accurately estimated.* Consider once again the IMPORT data. We have argued that there is an historical relationship between CONSUM and DOPROD that is approximated as CONSUM = (2/3)DOPROD. Replacing CONSUM in the original model we get

$$\text{IMPORT} = \beta_0 + (\beta_1 + (2/3)\beta_3) \cdot \text{DOPROD} + \beta_2 \cdot \text{STOCK} + u.$$

Equivalently stated, by dropping CONSUM from the equation we are able to obtain accurate estimates of $\beta_1 + (2/3)\beta_3$ and β_2. Multicollinearity is no longer present. The correlation between DOPROD and STOCK is .026. The results are given in Table 7.11. R^2 is almost unchanged and the residual plots (not shown) are satisfactory.

In this case we have used information in addition to the data to argue that the coefficient of DOPROD in the regression of IMPORT on DOPROD and STOCK is the linear combination $\beta_1 + (2/3)\beta_3$. Also, we have demonstrated that this linear function can be estimated accurately even though multicollinearity is present in the data. Whether or not it is useful to know the value of $\beta_1 + (2/3)\beta_3$ is, of course, another question. At least it is important to know that the estimate of the coefficient of DOPROD in this regression is not measuring the pure marginal effect of DOPROD, but includes part of the effect of CONSUM.

7.8.   THE PRINCIPAL COMPONENTS APPROACH

There is a constructive approach for identifying the linear combinations of the β's that can be accurately estimated. We shall use the data on sales, advertising, promotion, and sales expense (ADVER data set) introduced in the section 7.4 to demonstrate the method. The concepts are less intuitive

Table 7.11.   Estimate of a linear function for
IMPORT data

Linear function	Estimate	SE	t
$\beta_1 + (2/3)\beta_3$	.145	.007	20.70
β_2	.622	.128	4.86
$R^2 = .983$		$s = .667$	

*Refer to the Appendix to this chapter for further treatment of this problem.

than those found in the other sections of the chapter. We have attempted to keep things simple. A formal development of this problem is given in the Appendix to this chapter.

We begin with the linear transformation introduced in Section 7.5 that takes the standardized explanatory variables into a new orthogonal set of variables. The explanatory variables in the advertising data are advertising expenditure in the current year, promotion expenditure in the current year, sales expense, advertising expenditure lagged one year and promotion expenditure lagged one year. For simplicity of exposition in this section we refer to the standardized version of these variables as X_1 through X_5. The standardized variable, sales, is represented as Y. The transformation that takes $X_1, \ldots, X_5$ into the new set of orthogonal variables $Z_1, \ldots, Z_5$ is

$$Z_1 = -.532X_1 + .232X_2 + .389X_3 - .395X_4 + .595X_5, \qquad (7.9)$$

$$Z_2 = -.024X_1 + .825X_2 - .022X_3 - .260X_4 - .501X_5,$$

$$Z_3 = -.668X_1 + .158X_2 - .217X_3 + .692X_4 - .057X_5,$$

$$Z_4 = .074X_1 - .037X_2 + .895X_3 + .338X_4 - .279X_5,$$

$$Z_5 = .514X_1 + .489X_2 - .010X_3 + .428X_4 + .559X_5.$$

The coefficients in the equation defining Z_1 are the components of the characteristic vector corresponding to the largest characteristic root of the correlation matrix of the explanatory variables. Similarly, the coefficients defining Z_2 through Z_5 are components of the characteristic vectors corresponding to the remaining roots in order by size. The Z's are principal components as described in the previous section.

The regression model stated in terms of the original variables is [see Equation (7.4)]

$$S = \beta_0 + \beta_1 A + \beta_2 P + \beta_3 SE + \beta_4 A_{-1} + \beta_5 P_{-1} + u, \qquad (7.10)$$

where the subscript, -1, means that the variable is lagged by one period. In terms of standardized variables, the equation is written as

$$Y = \tilde{\beta}_1 X_1 + \tilde{\beta}_2 X_2 + \tilde{\beta}_3 X_3 + \tilde{\beta}_4 X_4 + \tilde{\beta}_5 X_5 + u'. \qquad (7.11)$$

The regression coefficients in Equation (7.11) are often referred to as beta coefficients. They represent marginal effects of the explanatory variables in standard deviation units. For example, $\tilde{\beta}_1$ measures the change in standardized units of sales corresponding to an increase of one standard deviation unit in A, advertising. Equation (7.11) has an equivalent form given as

$$Y = \alpha_1 Z_1 + \alpha_2 Z_2 + \alpha_3 Z_3 + \alpha_4 Z_4 + \alpha_5 Z_5 + u'. \qquad (7.12)$$

The equivalence of Equation (7.11) and Equation (7.12) results from the relationship between the X's and Z's in Equations (7.9) and the relationship between the α's and $\tilde{\beta}$'s and their estimated values, a's and $\tilde{b}$'s, given as

$$\tilde{b}_1 = -.532a_1 - .024a_2 - .668a_3 + .074a_4 + .514a_5, \qquad (7.13)$$

$$\tilde{b}_2 = .232a_1 + .825a_2 + .158a_3 - .037a_4 + .489a_5,$$

$$\tilde{b}_3 = .389a_1 - .022a_2 - .217a_3 + .895a_4 - .010a_5,$$

$$\tilde{b}_4 = -.395a_1 - .260a_2 + .692a_3 + .338a_4 + .428a_5,$$

$$\tilde{b}_5 = .595a_1 - .501a_2 - .057a_3 - .279a_4 + .559a_5.$$

Note that the transformation involves the same weights that are used to define (7.9). The advantage of the transformed model is that the Z's are orthogonal. The precision of the estimated regression coefficients as measured by variance of the a's is easily evaluated. The estimated variance of a_i is s^2/λ_i. It is inversely proportional to the ith characteristic root. All but α_5 may be accurately estimated since only λ_5 is small. (Recall that $\lambda_1 = 1.701$, $\lambda_2 = 1.288$, $\lambda_3 = 1.145$, $\lambda_4 = .859$, and $\lambda_5 = .007$.)

Our interest in the α's is only as a vehicle for analyzing the $\tilde{\beta}$'s. From the representation of Equation (7.13) it is a simple matter to compute and analyze the variances, and in turn, the standard errors, of the $\tilde{b}$'s. The estimated variance of any $\tilde{b}$ is the sum of the squares of the coefficients of the a's in Equation (7.13), multiplied by the variances of the a's. The sum is taken over a_1 through a_5. The estimated variance of a_i is $s^2/\lambda_i, i = 1, \ldots, 5$, where s^2 is the estimated variance of the residual term, u', in the model. For example, the variance of $\tilde{b}_1$ is

$$s_{\tilde{b}_1}^2 = \left(\frac{(-.532)^2}{\lambda_1} + \frac{(-.024)^2}{\lambda_2} + \frac{(-.668)^2}{\lambda_3} + \frac{(.074)^2}{\lambda_4} + \frac{(.514)^2}{\lambda_5} \right) s^2. \qquad (7.14)$$

Recall that $\lambda_1 \geqslant \lambda_2 \geqslant \cdots \geqslant \lambda_5$ and only λ_5 is small, $(\lambda_5 = .007)$. Therefore, it is only the last term in the expression for the variance that is large and could destroy the precision of $\tilde{b}_1$. Since expressions for the variances of the other $\tilde{b}$'s are similar to Equation (7.14), a requirement for small variance is equivalent to the requirement that the coefficient modifying $1/\lambda_5$ be small. Scanning the equations that define the transformation from $\{a_i\}$ to $\{\tilde{b}_i\}$ we see that $\tilde{b}_3$ is the most precise estimate since the coefficient of $1/\lambda_5$ in the variance expression for $\tilde{b}_3$ is $(.01)^2$ or $.0001$.

Expanding this type of analysis, it may be possible to identify meaningful linear functions of the $\tilde{\beta}$'s that can be more accurately estimated than individual $\tilde{\beta}$'s. For example, we may be more interested in estimating $\tilde{\beta}_1 - \tilde{\beta}_2$ than $\tilde{\beta}_1$ and $\tilde{\beta}_2$ separately. In the sales model, $\tilde{\beta}_1 - \tilde{\beta}_2$ measures the increment to sales that corresponds to increasing the current year's advertising budget, X_1, by one unit and simultaneously reducing the current year's promotions budget, X_2, by one unit. In other words, $\tilde{\beta}_1 - \tilde{\beta}_2$ represents the effect of a shift in the use of resources in the current year. The variance of $\tilde{b}_1 - \tilde{b}_2$, the estimate of $\tilde{\beta}_1 - \tilde{\beta}_2$, is obtained simply by subtracting the equation for $\tilde{b}_2$ from $\tilde{b}_1$ in (7.13) and using the resulting coefficients of the a's as before. That is

$$\tilde{b}_1 - \tilde{b}_2 = -.764a_1 - .849a_2 - .826a_3 + .111a_4 + .025a_5$$

and

$$s^2_{\tilde{b}_1 - \tilde{b}_2} = \left(\frac{(-.764)^2}{\lambda_1} + \frac{(-.849)^2}{\lambda_2} + \frac{(-.826)^2}{\lambda_3} + \frac{(.111)^2}{\lambda_4} + \frac{(.025)^2}{\lambda_5} \right) s^2.$$

$$(7.15)$$

The small coefficient modifying $1/\lambda_5$ makes it possible to accurately estimate $\tilde{\beta}_1 - \tilde{\beta}_2$. Generalizing this procedure we see that any linear function of the $\tilde{\beta}$'s that results in a small coefficient for $1/\lambda_5$ in the variance expression can be estimated with precision.

7.9. COMPUTATIONS ASSOCIATED WITH PRINCIPAL COMPONENTS

The computations required for this analysis involve something in addition to a standard least squares computer program. The raw data must be processed through a principal components subroutine that operates on the correlation matrix of the explanatory variables in order to compute the characteristic roots and the transformation weights found in Equations (7.13). Most regression packages produce the estimated beta coefficients as part of the standard output. These estimates are in column 1, labeled B (Std. V)* of Table 7.12. The estimate of $\tilde{\beta}_1 - \tilde{\beta}_2$ is $.583 - .973 = -.390$. It

---

*B (Std. V) means regression coefficient of a standardized variable.

should be noted that the t values for testing β_i and $\tilde{\beta}_i$ equal to zero are identical. The beta coefficient, $\tilde{\beta}_i$, is a scaled version of β_i. When constructing t values as either b_i/s_{b_i} or $\tilde{b}_i/s_{\tilde{b}_i}$, the scale factor is cancelled. The value of s^2 that appears in expressions (7.14) and (7.15) must reflect the scale factor. The correct value of s^2 is the value obtained from the original regression analysis, divided by the sum of squared deviations from the mean $(\Sigma(y_i - \bar{y})^2)$ of the response variable. In our current problem, $s^2 = 1.74243/335.451 = .0052$ (see Table 7.12). Alternatively, the estimated $\tilde{\beta}$'s and the value of s^2 can be obtained directly from a regression of the standardized variables as represented in Equation (7.11). By either method, the variance of $\tilde{b}_1 - \tilde{b}_2$ can be computed from Equation (7.15) as .0083. A 95% confidence interval for $\tilde{\beta}_1 - \tilde{\beta}_2$ is $-.390 \pm (2.12)(.0083)^{1/2}$ or $-.583$ to $-.197$. That is, the effect of shifting one unit of expenditure from promotions to advertising in the current year is a loss of between .197 and .583 standardized sales units.

There are other linear functions that may also be accurately estimated. Any function that produces a small coefficient for $1/\lambda_5$ in the variance expression is a possibility. For example, Equations (7.12) suggest that all differences involving $\tilde{b}_1$, $\tilde{b}_2$, $\tilde{b}_4$, and $\tilde{b}_5$ can be considered. However, some of the differences are meaningful in the problem whereas others are not. For example, the difference $(\tilde{\beta}_1 - \tilde{\beta}_2)$ is meaningful as described previously. It represents a shift in current expenditures from promotions to advertising. The difference $\tilde{\beta}_1 - \tilde{\beta}_4$ is not particularly meaningful. It represents a shift from current advertising expenditure to a previous year's advertising expenditure. A shift of resources backward in time is impossible. Even though $\tilde{\beta}_1 - \tilde{\beta}_4$ could be accurately estimated, it is not of interest in the analysis of sales.

In general, when the weights in Equations (7.13) are displayed and the corresponding values of the characteristic roots are known, it is always possible to scan the weights and identify those linear functions of the original regression coefficients that can be accurately estimated. Of those

Table 7.12. Regression results; sales, advertising, promotion Data: Standardized variables

Variable	B(Std. V)	SE	t
A	.5830	.4380	1.331
P	.9734	.4170	2.334
SE	.7859	.07476	10.512
A_{-1}	.3953	.3670	1.077
P_{-1}	.5035	.4754	1.059
$n = 22$	$R^2 = .9169$	$s = .0721$	

linear functions that can be accurately estimated, only some will be of interest for the problem being studied.

To summarize, where multicollinearity is indicated and it is not possible to supplement the data, it may still be possible to estimate some regression coefficients and some linear functions accurately. To investigate which coefficients and linear functions can be estimated, we recommend the analysis (transformation to Principal Components) that has just been described. This method of analysis will not overcome multicollinearity if it is present. There will still be regression coefficients and functions of regression coefficients that cannot be estimated. But the recommended analysis will indicate those functions that are estimable and indicate the structural dependencies which exist among the explanatory variables.

BIBLIOGRAPHIC NOTES

The principal components techniques that are used in this chapter are derived in most books on multivariate statistical analysis. It should be noted that principal components analysis involves only the explanatory variables. The analysis is aimed at characterizing the nature of the explanatory information and indentifying dependencies (if they exist) among the explanatory variables. The principal components problem is thoroughly discussed in *Applied Multivariate Analysis* (Press, 1972).

REFERENCES

Coleman, J. S., E. Q. Cambell, C. J. Hobson, J. McPartland, A. M. Mood, F. D. Weinfield, and R. L. York, *Equality of Educational Opportunity*, Washington D. C., U.S. Government Printing Office, 1966.

Kendall, M. G., *A Course in Multivariate Analysis*, Charles Griffin, London, 1957.

Malinvaud, E., *Statistical Methods of Econometrics*, Rand-McNally, Chicago, 1968.

Mosteller, F. and D. P. Moynihan, Eds., *On Equality of Educational Opportunity*, Random House, New York, 1972.

Press, S. J., *Applied Multivariate Analysis*, Holt, Rinehart, and Winston, New York, 1972.

Silvey, S. D., Multicollinearity and imprecise estimation, *J. Royal Stat. Soc.*, **31**, 539–552 (1969).

APPENDIX: PRINCIPAL COMPONENTS

THE MODEL

The regression model is

$$Y = X\beta + u \tag{1}$$

where $\mathbf{Y}$ is an $n \times 1$ vector of observations on the response variable, $\mathbf{X} = (\mathbf{X}_{(1)}, \ldots, \mathbf{X}_{(p)})$ is an $n \times p$ matrix of n observations on p explanatory variables, $\boldsymbol{\beta}$ is a $p \times 1$ vector of regression coefficients and $\mathbf{u}$ is an $n \times 1$ vector of residuals. It is assumed that $E(\mathbf{u}) = \mathbf{0}$, $E(\mathbf{u}\mathbf{u}') = \sigma^2 \mathbf{I}$ and that $\mathbf{X}$ and $\mathbf{Y}$ have been centered and scaled so that $\mathbf{X}'\mathbf{X}$ and $\mathbf{X}'\mathbf{Y}$ are matrices of correlation coefficients.

There exists a matrix, $\mathbf{C}$, satisfying

$$\mathbf{C}'(\mathbf{X}'\mathbf{X})\mathbf{C} = \boldsymbol{\Lambda} \quad \text{and} \quad \mathbf{C}'\mathbf{C} = \mathbf{C}\mathbf{C}' = \mathbf{I}, \tag{2}$$

where $\boldsymbol{\Lambda}$ is a diagonal matrix with the ordered characteristic roots of $\mathbf{X}'\mathbf{X}$ on the diagonal. The characteristic roots are denoted by $\lambda_1 \geqslant \lambda_2 \geqslant \lambda_3 \cdots \geqslant \lambda_p$. The columns of $\mathbf{C}$ are the normalized characteristic vectors corresponding to $\lambda_1, \ldots, \lambda_p$. $\mathbf{C}$ may be used to calculate a new set of explanatory variables, namely

$$(\mathbf{W}_{(1)}, \mathbf{W}_{(2)}, \ldots, \mathbf{W}_{(p)}) = \mathbf{W} = \mathbf{X}\mathbf{C} = (\mathbf{X}_{(1)}, \ldots, \mathbf{X}_{(p)})\mathbf{C} \tag{3}$$

that are linear functions of the original explanatory variables. The $\mathbf{W}$'s are referred to as principal components. The regression model of Equation (1) can be restated in terms of the principal components as

$$\mathbf{Y} = \mathbf{W}\boldsymbol{\alpha} + \mathbf{u}, \tag{4}$$

where $\mathbf{W} = \mathbf{X}\mathbf{C}$ and $\boldsymbol{\alpha} = \mathbf{C}'\boldsymbol{\beta}$.

COLLINEAR DATA

The principal components and the characteristic roots may be used to detect and analyze collinearity in the explanatory variables. The restatement of the regression model given in Equation (4) is a reparameterization of Equation (1) in terms of orthogonal explanatory variables. That is, $\mathbf{W}'_{(i)}\mathbf{W}_{(j)} = 0$ for $i \neq j$ and $\mathbf{W}'_{(i)}\mathbf{W}_{(i)} = \lambda_i$. The λ's may be viewed as sample variances of the principal components. If $\lambda_i = 0$, then all observations on the ith principal component are also zero. Since the ith principal component is a linear function of the X's, it follows therefore, that when $\lambda_i = 0$ an exact linear dependence exists among the explanatory variables. It follows also that when λ_i is small (approximately equal to zero) there is an approximate linear relationship among the explanatory variables. That is, a small characteristic root is an indicator of multicollinearity. In addition, from Equation (3) we have

$$\mathbf{W}_{(i)} = \sum_{k=1}^{p} C_{ki}\mathbf{X}_{(k)}$$

which identifies the exact form of the linear relationship that is causing the multicollinearity.

PRECISION OF LINEAR FUNCTIONS OF $\hat{\beta}$

Denoting $\hat{\alpha}$ and $\hat{\beta}$ as the least squares estimators for α and β, respectively, it can be shown that

$$\hat{\alpha} = C'\hat{\beta}, \text{ and conversely } \hat{\beta} = C\hat{\alpha}.$$

With $\hat{\alpha} = (W'W)^{-1}W'Y$, it follows that the variance–covariance matrix of $\hat{\alpha}$ is

$$V(\hat{\alpha}) = \Lambda^{-1}\sigma^2,$$

and the corresponding matrix for $\hat{\beta}$ is

$$V(\hat{\beta}) = C\Lambda^{-1}C'\sigma^2.$$

Let L be an arbitrary $p \times 1$ vector of constants. The linear function $\varphi = L'\beta$ has least squares estimator $\hat{\varphi} = L'\hat{\beta}$ and variance

$$\text{Var}(\hat{\varphi}) = L'C\Lambda^{-1}C'L\sigma^2. \tag{5}$$

Let $C_{(i)}$ be the ith column of C. L can be represented as $L = \sum_{i=1}^{p} r_i C_{(i)}$ for appropriately chosen constants $r_1, \ldots, r_p$. Then Equation (5) becomes $\text{Var}(\hat{\varphi}) = R'\Lambda^{-1}R\sigma^2$ or

$$\text{Var}(\hat{\varphi}) = \left(\sum \frac{r_i^2}{\lambda_i} \right)\sigma^2. \tag{6}$$

To summarize, the variance of $\hat{\varphi}$ is a simple linear combination of the reciprocals of the characteristic roots. It follows that $\hat{\varphi}$ will have good precision if either none of the characteristic roots are near zero or if r_i^2 is at most the same magnitude as λ_i when λ_i is small. Furthermore, it is always possible to select a vector, L, and thereby a linear function of $\hat{\beta}$ so that the effect of one or two small characteristic roots is eliminated and $L'\hat{\beta}$ has a small variance. Refer to Silvey (1969) for a more complete development of these concepts.

CHAPTER 8

Biased Estimation of Regression Coefficients

8.1. INTRODUCTION

It was demonstrated in the last chapter that when multicollinearity is present in a set of explanatory variables, the ordinary least squares estimates of the individual regression coefficients tend to be unstable and can lead to erroneous inferences. In this chapter, two alternative estimation methods that provide a more informative analysis of the data than the OLS method when multicollinearity is present are considered. The estimators discussed here are biased, but tend to have more precision (as measured by mean square error) than the OLS estimators. (See Kendall, 1957; McCallum, 1970; and Hoerl and Kennard, 1970.) These alternative methods do not reproduce the estimation data as well as the OLS method; the sum of squared residuals is not as small and, equivalently, the multiple correlation coefficient is not as large. However, the two alternatives have the potential to produce more precision in the estimated coefficients and smaller prediction errors when the predictions are generated using data other than those used for estimation.

Unfortunately, the criteria for deciding when these methods give better results than the OLS method depend on the true but unknown values of the model regression coefficients. That is, there is no completely objective way to decide when OLS should be replaced in favor of one of the alternatives. Nevertheless, when multicollinearity is suspected, the alternative methods of analysis are recommended. The resulting estimated regression coefficients may suggest a new interpretation of the data that, in turn, can lead to a better understanding of the process under study.

The two specific alternatives to OLS that are considered are (i) principal components regression and (ii) ridge regression. Principal components analysis was introduced in Chapter 7. It is assumed that the reader is

175

familiar with that material. It will be demonstrated that the principal components estimation method can be interpreted in two ways; one interpretation relates to the nonorthogonality of the explanatory variables, the other has to do with constraints on the regression coefficients. Ridge regression also involves constraints on the coefficients. The ridge method is introduced in this chapter and it is applied again in Chapter 9 to the problem of variable selection. Both methods, principal components and ridge regression, are examined using the French import data that was analyzed in Chapter 7.

8.2. PRINCIPAL COMPONENTS REGRESSION

The model under consideration is

$$\text{IMPORT} = \beta_0 + \beta_1 \cdot \text{DOPROD} + \beta_2 \cdot \text{STOCK} + \beta_3 \cdot \text{CONSUM} + u. \quad (8.1)$$

The variables are defined in Chapter 7, p. 151. The principal components of the standardized explanatory variables are (see Equation (7.5))

$$Z_1 = .7063 \cdot \text{doprod} + .0435 \cdot \text{stock} + .7065 \cdot \text{consum},$$

$$Z_2 = -.0357 \cdot \text{doprod} + .9990 \cdot \text{stock} - .0258 \cdot \text{consum}, \quad (8.2)$$

$$Z_3 = -.7070 \cdot \text{doprod} - .0070 \cdot \text{stock} + .7072 \cdot \text{consum}.$$

The Z's have sample variances $\lambda_1 = 1.999$, $\lambda_2 = .998$, and $\lambda_3 = .003$, respectively. Recall that the λ's are the characteristic roots of the correlation matrix of DOPROD, STOCK, and CONSUM. The model of Equation (8.1) stated in terms of standardized variables is

$$\text{import} = \tilde{\beta}_1 \cdot \text{doprod} + \tilde{\beta}_2 \cdot \text{stock} + \tilde{\beta}_3 \cdot \text{consum} + u. \quad (8.1a)$$

It may be written in terms of the principal components as

$$\text{import} = \alpha_1 Z_1 + \alpha_2 Z_2 + \alpha_3 Z_3 + u. \quad (8.3)$$

Equation (8.3) is equivalent to Equation (8.1a). In addition, the Z's are orthogonal. The equivalence follows since there is a unique relationship between the α's and $\tilde{\beta}$'s. In particular,

$$\alpha_1 = .7063 \tilde{\beta}_1 + .0435 \tilde{\beta}_2 + .7065 \tilde{\beta}_3,$$

$$\alpha_2 = -.0357 \tilde{\beta}_1 + .9990 \tilde{\beta}_2 - .0258 \tilde{\beta}_3, \quad (8.4)$$

$$\alpha_3 = -.7070 \tilde{\beta}_1 - .0070 \tilde{\beta}_2 + .7072 \tilde{\beta}_3.$$

Conversely,

$$\tilde{\beta}_1 = .7063\alpha_1 - .0357\alpha_2 - .7070\alpha_3,$$

$$\tilde{\beta}_2 = .0435\alpha_1 + .9990\alpha_2 - .0070\alpha_3, \qquad (8.5)$$

$$\tilde{\beta}_3 = .7065\alpha_1 - .0258\alpha_2 + .7072\alpha_3.$$

These same relationships hold for the least squares estimates, the a_i's and $\tilde{b}_i$'s of the α's and $\tilde{\beta}$'s, respectively. Therefore, the $\tilde{a}_i$'s and $\tilde{b}_i$'s may be obtained by the regression of import against the principal components Z_1, Z_2, and Z_3, or against the original standardized variables.

Observe that the regression relationship given in terms of the principal components (Equation (8.3)) is not easily interpreted. The explanatory variables of that model are linear combinations of the original explanatory variables. The α's, unlike the $\tilde{\beta}$'s, do not have simple interpretations as marginal effects of the original explanatory variables. Therefore, we use principal components regression only as a means for analyzing the multicollinearity problem. The final estimation results are always restated in terms of the $\tilde{\beta}$'s for interpretation.

8.3.   REMOVING DEPENDENCE AMONG THE EXPLANATORY VARIABLES

It has been mentioned that the principal components regression has two interpretations. We shall first use the principal components technique to reduce multicollinearity in the estimation data. The reduction is accomplished by using less than the full set of principal components to explain the variation in the response variable. (When all three principal components are used, the OLS solution is exactly reproduced by applying Equations (8.5).)

Since Z_3 has variance equal to .003, the linear function defining Z_3 is approximately equal to zero and is the source of multicollinearity in the data. We shall exclude Z_3 and consider regressions of import against Z_1 alone as well as against Z_1 and Z_2. We consider the two possible regression models

$$\text{import} = \alpha_1 Z_1 + u \qquad (8.6)$$

and

$$\text{import} = \alpha_1 Z_1 + \alpha_2 Z_2 + u. \qquad (8.7)$$

Both models lead to estimates for all three of the original coefficients, $\tilde{\beta}_1$,

$\tilde{\beta}_2$, and $\tilde{\beta}_3$. The estimates are biased since some information (Z_3 in Equation (8.7), Z_2 and Z_3 in Equation (8.6)) has been excluded in both cases.

The estimated values of α_1 or α_1 and α_2 may be obtained by regressing import in turn against Z_1 and then against Z_1 and Z_2. However, a simpler computational method is available that exploits the orthogonality of Z_1, Z_2, and Z_3.* For example, the same estimated value of α_1 will be obtained from regression using (8.3), (8.6), or (8.7). Similarly, the value of α_2 may be obtained from (8.3) or (8.7). It also follows that if we have the OLS estimates of the $\tilde{\beta}$'s, estimates of the α's may be obtained from Equations (8.4). Then principal components regression estimates of the $\tilde{\beta}$'s corresponding to (8.6) and (8.7) can be computed by referring back to Equations (8.5) and setting the appropriate α's to zero. The following examples will clarify the process.

Regression estimates for the import relationship are given in Table 8.1.** From Equation (8.4), the corresponding estimates of the α's are .6900, .1913, 1.1597. Using $\alpha_1 = .6900$ and $\alpha_2 = \alpha_3 = 0$ in Equations (8.5) yields estimated $\tilde{\beta}$'s corresponding to regression on only the first principal component. (Results are in Table 8.2.) Equivalently, the definition of Z_1 from (8.2) may be used in (8.6) to obtain

$$\text{import} = .6900(.7063 \cdot \text{doprod} + .0435 \cdot \text{stock} + .7065 \cdot \text{consum})$$

$$= .4873 \cdot \text{doprod} + .0300 \cdot \text{stock} + .4875 \cdot \text{consum}.$$

Estimated $\tilde{\beta}$'s are obtained for regression on the first two principal components in a similar fashion, $\alpha_1 = .6900$, $\alpha_2 = .1913$, and $\alpha_3 = 0$ in (8.5).

From Table 8.2 it is evident that using different numbers of principal

Table 8.1.   IMPORT *data:* 1949–1959

Variable	B(Std. V)	Coefficient	SE	t
DOPROD	−.3394	−.0514	.0703	−.731
STOCK	.2130	.5869	.0046	6.203
CONSUM	1.3028	.2868	.1022	2.806
CONSTANT	0	−10.1300	1.2122	−8.355

*In any regression equation where the full set of potential explanatory variables under consideration are orthogonal, the estimated values of regression coefficients are not altered when subsets of these variables are either introduced or deleted.

**Most regression computer packages produce values for both the regular and standardized regression coefficients. The estimated coefficients satisfy $b_i = \tilde{b}_i(s_y/s_{x_i})$ where s_y and s_{x_i} are standard deviations of the response and ith explanatory variable, respectively.

Table 8.2.   *Principal components regression results for* IMPORT *data*

Variable	First principal component Equation (8.5)		First and second principal components Equation (8.7)		All principal components—OLS Equation (8.1)	
	$\tilde{b}$	b	$\tilde{b}$	b	$\tilde{b}$	b
DOPROD	.4873	.0738	.4804	.0727	−.3394	−.0514
STOCK	.0300	.0826	.2211	.6091	.2130	.5869
CONSUM	.4875	.1073	.4825	.1062	1.3028	.2868
CONSTANT	0	−7.7350	0	−9.1057	0	−10.1300
R^2		.952		.988		.992

components gives substantially different results. It has already been argued that the OLS estimates are unsatisfactory. The negative coefficient of doprod is unexpected and cannot be sensibly interpreted. Furthermore, there is extensive multicollinearity which enters through the principal component, Z_3. This variable has almost zero variance ($\lambda_3 = .003$) and is therefore approximately equal to zero. Of the two remaining principal components, it is fairly clear that the first one is associated with the combined effect of DOPROD and CONSUM. The second principal component is uniquely associated with STOCK. This conclusion is apparent in Table 8.2. The coefficients of DOPROD and CONSUM are completely determined from the regression of IMPORT on Z_1 alone. These coefficients do not change when Z_2 is used. The addition of Z_2 causes the coefficient of STOCK to increase from .0826 to .6091. Also, R^2 increases from .952 to .988. Selecting the model based on the first two principal components, the resulting equation stated in original units is

$$\text{IMPORT} = -9.1057 + .0727 \cdot \text{DOPROD}$$
$$+ .6091 \cdot \text{STOCK} + .1062 \cdot \text{CONSUM}. \qquad (8.8)$$

It provides a different and more plausible representation of the IMPORT relationship than was obtained from the OLS results. In addition, the analysis has led to an explicit quantification (in standardized variables) of the linear dependency in the explanatory variables. We have $Z_3 = 0$ or equivalently [from Equations (8.2)]

$$-.7070 \cdot \text{doprod} - .0070 \cdot \text{stock} + .7072 \cdot \text{consum} \doteq 0.$$

The standardized values of DOPROD and CONSUM are essentially equal. This information can be useful qualitatively and quantitatively if Equation (8.8) is used for forecasting or for analyzing policy decisions.

8.4. CONSTRAINTS ON THE REGRESSION COEFFICIENTS

There is a second interpretation of the results of the principal components regression equation. The interpretation is linked to the notion of imposing constraints on the $\tilde{\beta}$'s which was introduced in the previous chapter. The estimates for Equation (8.7) were obtained by setting α_3 equal to zero in Equations (8.5). From (8.4), $\alpha_3 = 0$ implies that

$$-.7070\tilde{\beta}_1 - .0070\tilde{\beta}_2 + .7072\tilde{\beta}_3 = 0 \qquad (8.9)$$

or

$$\tilde{\beta}_1 \doteq \tilde{\beta}_3.$$

In original units, Equation (8.9) becomes

$$-6.67\beta_1 + 4.54\beta_3 = 0 \qquad (8.10)$$

or

$$\beta_1 = .6872\beta_3.$$

Therefore, the estimates obtained by regression on Z_1 and Z_2 could have been obtained using OLS as in the preceding chapter with a linear constraint on the coefficients given by Equation (8.10).

Recall that in Chapter 7 we conjectured that $\beta_1 = \beta_3$ as a prior constraint on the coefficients. It was argued that the constraint was the result of a qualitative judgment based on knowledge of the process under study. It was imposed without looking at the data. Now, using the data, we have found that principal components regression on Z_1 and Z_2 gives a result that is equivalent to imposing the constraint of Equation (8.10). The result suggests that the marginal effect of domestic production on imports is about 69% of the marginal effect of domestic consumption on imports.

To summarize, the method of principal components regression provides both alternative estimates of the regression coefficients as well as other useful information about the underlying process that is generating the data. The structure of linear dependence among the explanatory variables is made explicit. Principal components with small variances (characteristic roots) exhibit the linear relationships among the original variables that are the source of multicollinearity. Also elimination of multicollinearity by dropping one or more principal components from the regression is equivalent to imposing constraints on the regression coefficients. It provides a constructive way of identifying those constraints that are consistent with the proposed model and the information contained in the data.

8.5.   RIDGE REGRESSION*

Ridge regression provides another alternative estimation method that may be used to advantage when the explanatory variables are highly nonorthogonal. There are a number of alternative ways to define and compute ridge estimates. (See the Appendix to this chapter.) We have chosen to present the method associated with the ridge trace. It is a graphical approach and may be viewed as an exploratory technique. Ridge analysis using the ridge trace represents a unified approach to problems of detection and estimation when multicollinearity is suspected. The estimators produced are biased, but tend to have a smaller mean square error than OLS estimators (Hoerl and Kennard, 1970). The ridge estimates are stable in the sense that they are not affected by slight variations in the estimation data. Because of the smaller mean square error property, values of the ridge estimated coefficients are expected to be closer than the OLS estimates to the true values of the regression coefficients. Also, forecasts of the response variable corresponding to values of the explanatory variables not included in the estimation set tend to be more accurate.

As with the principal components method, the criteria for deciding when the ridge estimators have an important advantage over the OLS estimators depend on the values of the true regression coefficients in the model. Although these values cannot be known, we still suggest that ridge analysis is useful in cases where extreme multicollinearity is suspected. The ridge coefficients can suggest an alternative interpretation of the data that may lead to a better understanding of the process under study.

8.6.   DEFINITION AND COMPUTATION**

Ridge estimates of the regression coefficients may be obtained by solving a slightly altered form of the normal equations. (The normal equations are introduced in Chapter 3.) Assume that the standardized form of the regression model is given as

$$y = \beta_1 x_1 + \beta_2 x_2 + \cdots + \beta_p x_p + u. \tag{8.11}$$

The estimating equations for the ridge regression coefficients are

$$(1+k)\tilde{\beta}_1 + r_{12}\tilde{\beta}_2 + \cdots + r_{1p}\tilde{\beta}_p = r_{1y},$$

$$r_{12}\tilde{\beta}_1 + (1+k)\tilde{\beta}_2 + \cdots + r_{2p}\tilde{\beta}_p = r_{2y}, \tag{8.12}$$

$$r_{1p}\tilde{\beta}_1 + r_{2p}\tilde{\beta}_2 + \cdots + (1+k)\tilde{\beta}_p = r_{py},$$

*Hoerl (1959) named the method ridge regression because of its similarity to ridge analysis used in his earlier work to study second-order response surfaces in many variables.
**See the Appendix for a formal treatment.

where r_{ij} is the simple correlation between the ith and jth explanatory variables and r_{iy} is the correlation between the ith explanatory variable and the response variable y. The solution to (8.12), $\tilde{b}_1, \ldots, \tilde{b}_p$, is the set of estimated ridge regression coefficients. The essential parameter that distinguishes ridge regression from OLS is k. Note that when $k = 0$, the $\tilde{b}$'s are the OLS estimates. The parameter k may be referred to as the bias parameter. As k increases from zero, bias of the estimates increases. As k continues to increase without bound, the regression estimates all tend toward zero. It has been shown that there is a positive value of k for which the ridge estimates will be stable with respect to small changes in the estimation data, (Hoerl and Kennard, 1970). In practice, a value of k is chosen by computing $\tilde{b}_1, \ldots, \tilde{b}_p$ for a range of k values between 0 and 1 and plotting the results against k. The resulting graph is known as the ridge trace and is used to select an appropriate value for k. Guidelines for choosing k are given in the following example.

8.7. DETECTION OF MULTICOLLINEARITY USING RIDGE METHODS

There are two methods for detecting multicollinearity that are inherently associated with ridge regression. The first method is related to the effect that multicollinearity has on the error between the OLS estimators and the true values of the regression coefficients. The second method deals with the basic instability that OLS estimates exhibit in response to slight changes in the estimation data.

The first method is associated with the notion of a variance inflation factor. The precision of an OLS estimated regression coefficient is measured by its variance which is proportional to σ^2, the variance of the residual term in the regression model. The constant of proportionality may be referred to as the variance inflation factor, VIF.* There is a VIF corresponding to each OLS estimated coefficient. The VIF for $\tilde{b}_i$ is equal to $(1 - R_i^2)^{-1}$, where R_i^2 is the square of the multiple correlation coefficient from the regression of the ith explanatory variable on all other explanatory variables in the equation. As R_i^2 tends toward 1 indicating the presence of a linear relationship in the x's, the VIF for the estimated coefficient of x_i tends to infinity. It is suggested that a VIF in excess of 10 is an indication that multicollinearity may be causing problems in estimation.

The VIF's may be used to obtain an expression for the expected squared distance of the OLS estimators from their true values. This distance is another measure of precision of the least squares estimators. The smaller

---

*VIF is the diagonal element of the inverse of correlation matrix. See Appendix.

the distance, the more accurate are the estimates. Denoting the square of the distance by L^2, it follows, that on average (see Appendix),

$$L^2 = \sigma^2 \sum_{i=1}^{p} \text{VIF}_i.$$

If the explanatory variables were orthogonal, the VIF's would all be equal to 1. Then L^2 would take the value $p\sigma^2$. It follows that the ratio

$$R_L = \frac{\sigma^2 \sum_{i=1}^{p} \text{VIF}_i}{p\sigma^2} = \frac{\sum_{i=1}^{p} \text{VIF}_i}{p},$$

which measures the squared error in the OLS estimators relative to the size of that error if the data were orthogonal, may also be used as an index of multicollinearity.

Using the IMPORT data once again as the example, the VIF's for the coefficients of DOPROD, STOCK, and CONSUM are 185.99, 1.02, and 186.10, respectively. R_L for this data is 124.37. That is, the squared error in the OLS estimators is 124 times as large as it would be if the explanatory variables were orthogonal. The size of R_L points to multicollinearity.

The use of R_L and the VIF's is probably more than what is required to detect multicollinearity in the French import data. Recall that the simple correlation between DOPROD and CONSUM is .997. However, the example is good for illustrative purposes. The methods are applied to a data set with a more complex collinear structure in Chapter 9.

The second method for detecting multicollinearity that comes out of ridge analysis deals with the instability in the estimated coefficients resulting from slight changes in the estimation data. The instability may be observed in the ridge trace. The ridge trace is a simultaneous graph of the p regression coefficients plotted against k. We have described k as the bias parameter. In addition, for small values, $k = .001, .002$, and so on, the corresponding ridge estimates may be viewed as resulting from a set of data that has been slightly altered. Figure 8.1 is the ridge trace for the IMPORT data. The graph is constructed from Table 8.3 which has the ridge estimated coefficients for 29 values of k ranging from 0 to 1. Typically, the values of k are chosen to be concentrated near the low end of the range. If the estimated coefficients show large fluctuations for small values of k, then instability has been demonstrated and multicollinearity is probably at work.

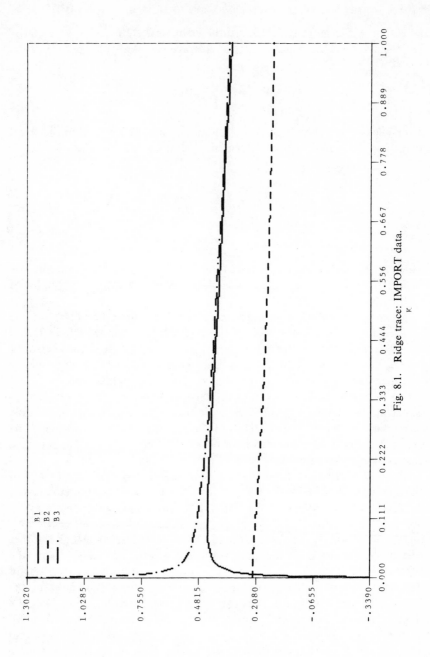

Fig. 8.1. Ridge trace: IMPORT data.

κ

184

Table 8.3.   Ridge estimates: IMPORT data

	K	B1	B2	B3
001	0.000	-.339	0.213	1.302
002	0.001	-.117	0.215	1.080
003	0.002	0.010	0.216	0.952
004	0.003	0.092	0.217	0.870
005	0.004	0.150	0.217	0.811
006	0.005	0.193	0.217	0.768
007	0.006	0.225	0.217	0.735
008	0.007	0.251	0.217	0.709
009	0.008	0.272	0.217	0.687
010	0.009	0.290	0.217	0.669
011	0.010	0.304	0.217	0.654
012	0.020	0.379	0.217	0.575
013	0.030	0.406	0.214	0.543
014	0.040	0.420	0.213	0.525
015	0.050	0.427	0.211	0.513
016	0.060	0.432	0.209	0.504
017	0.070	0.434	0.207	0.497
018	0.080	0.436	0.206	0.491
019	0.090	0.436	0.204	0.486
020	0.100	0.436	0.202	0.481
021	0.200	0.426	0.186	0.450
022	0.300	0.411	0.173	0.427
023	0.400	0.396	0.161	0.408
024	0.500	0.381	0.151	0.391
025	0.600	0.367	0.142	0.376
026	0.700	0.354	0.135	0.361
027	0.800	0.342	0.128	0.348
028	0.900	0.330	0.121	0.336
029	1.000	0.319	0.115	0.325

8.8.   ESTIMATION BY THE RIDGE METHOD

What is evident from the trace or equivalently from Table 8.3 is that the estimated values of the coefficients for $\tilde{\beta}_1$ and $\tilde{\beta}_3$ are quite unstable for small values of k. The estimate of $\tilde{\beta}_1$ changes rapidly from an implausible negative value of $-.3394$ to a stable value of about .42. The estimate of $\tilde{\beta}_3$ retreats from 1.3028 to stabilize at about .52. The coefficient of stock, $\tilde{\beta}_2$, is unaffected by the multicollinearity and remains stable throughout at about .21.

The next step in the ridge analysis is to select a value of k and the corresponding estimates of the regression coefficients. Guides to selecting k have been outlined by Hoerl and Kennard (1970). The guides are reproduced as follows:

1.   At a certain value of k, the system will stabilize and have the general character of an orthogonal system.

2.   Coefficients will not have unreasonable absolute values with respect to factors for which they represent rates of change.

3.   Coefficients with improper signs at $k=0$ will have changed to have proper signs.

4. The residual sum of squares will not have been inflated to an unreasonable value. It will not be large relative to the minimum residual sum of squares or large relative to what would be a reasonable variance for the process generating the data.

For the IMPORT data, the ridge trace appears to stabilize for k slightly larger then .01. The VIF's get down to around unity (Table 8.4), a value characteristic of an orthogonal system, for $k = .03$ or .04. We select $k = .04$ as the solution. It is desirable to select the smallest value of k for which stability occurs since the size of k is directly related to the amount of bias introduced. At $k = .04$, the improper negative sign on the estimate of $\tilde{\beta}_1$ has disappeared and the coefficient has stabilized at about .42. The sum of squared residuals has only increased from .0081 at $k = 0$ to .0117 and R^2 has gone from .992 to .988. The solution at $k = .04$ appears to be satisfactory on all counts. The results are summarized in Table 8.5.

Table 8.4.   Variance inflation factors (VIF) for regression coefficients: IM-PORT *data*

ROW	K	B 1	B 2	B 3
* 1 *	0.000	185.941	1.019	186.059
* 2 *	0.001	98.972	1.008	99.030
* 3 *	0.002	61.369	1.003	61.405
* 4 *	0.003	41.779	0.999	41.804
* 5 *	0.004	30.295	0.996	30.313
* 6 *	0.005	22.989	0.993	23.002
* 7 *	0.006	18.058	0.991	18.068
* 8 *	0.007	14.572	0.988	14.581
* 9 *	0.008	12.018	0.986	12.025
* 10 *	0.009	10.090	0.984	10.096
* 11 *	0.010	8.601	0.982	8.605
* 12 *	0.020	2.859	0.962	2.860
* 13 *	0.030	1.502	0.943	1.503
* 14 *	0.040	0.979	0.925	0.979
* 15 *	0.050	0.723	0.908	0.723
* 16 *	0.060	0.579	0.891	0.578
* 17 *	0.070	0.489	0.874	0.488
* 18 *	0.080	0.429	0.858	0.428
* 19 *	0.090	0.386	0.842	0.386
* 20 *	0.100	0.355	0.827	0.355
* 21 *	0.200	0.240	0.695	0.240
* 22 *	0.300	0.204	0.592	0.204
* 23 *	0.400	0.182	0.510	0.182
* 24 *	0.500	0.166	0.444	0.165
* 25 *	0.600	0.152	0.391	0.152
* 26 *	0.700	0.140	0.346	0.140
* 27 *	0.800	0.130	0.309	0.130
* 28 *	0.900	0.121	0.277	0.121
* 29 *	1.000	0.113	0.250	0.112

Table 8.5.   **Ridge estimates**[a] **for** IMPORT **data**

Variable	OLS $k=0$		Ridge $k=.04$	
	$\tilde{b}$	b	$\tilde{b}$	b
DOPROD	$-.3394$	$-.0514$	.4196	.0635
STOCK	.2130	.5869	.2127	.5859
CONSUM	1.3028	.2868	.5249	.1156
CONSTANT	0	-10.1300	0	-8.5537
R^2		.992		.988

[a]The table displays estimated coefficients from the model stated in standardized variables $\tilde{b}$, and in original units b. Recall that if the model is $y = \beta_0 + \beta_1 x_1 + \cdots + \beta_p x_p + u$ in original variables, $b_i = \tilde{b}_i \ (s_y / s_{x_i})$.

The resulting model for $k = .04$ is

$$\text{IMPORT} = -8.5537 + .0635 \cdot \text{DOPROD}$$
$$+ .5859 \cdot \text{STOCK} + .1156 \cdot \text{CONSUM}.$$

The equation gives a plausible representation of the relationship. Note that the final equation for this data is not particularly different from the result obtained by using the first two principal components (see Table 8.2), although the two computational methods appear to be very different.

8.9. SUMMARY

Both alternative estimation methods, ridge regression and principal components regression, provide additional information about the data being analyzed. We have seen that the characteristic roots of the correlation matrix of explanatory variables play an important role in detecting multicollinearity and in analyzing its effects. The regression estimates produced by these methods are biased, but may be more accurate than OLS estimates in terms of mean square error. It is impossible to evaluate the gain in accuracy for a specific problem since a comparison of the two methods to OLS requires knowledge of the true values of the coefficients. Nevertheless, when severe multicollinearity is suspected, it is worth the effort to produce at least one set of estimates in addition to the OLS estimates. The estimates may suggest an interpretation of the data that was not previously considered.

BIBLIOGRAPHIC NOTES

Ridge regression is a relatively new technique and, as such, its full value as a practical tool has not yet been adequately investigated. There are a number of empirical studies that begin to provide the type of insight that a practitioner needs. For example, see Dempster, Schatzoff, and Wermuth (1976). A fairly complete list of references to work on ridge regression can be found in Hocking (1976).

REFERENCES

Dempster, A. P., M. Schatzoff, and N. Wermuth, A simulation study of alternatives to ordinary least squares, Research Report S–35, Department of Statistics, Harvard University, 1975.

Hocking, R. R., The analysis and selection of variables in linear regression, *Biometrics*, **32**, 1–49 (1976).

Hoerl, A. E., Optimum solution of many variables, *Chem. Eng. Q. Progr.*, **55**, 69–78 (1959).

Hoerl, A. E. and R. W. Kennard, Ridge regression: Biased estimation for nonorthogonal problems, *Technometrics*, **12**, 69–82 (1970).

Kendall, M. G., *A Course in Multivariate Analysis*, Charles Griffin, London, 1957.

McCallum, B. T., Artificial orthogonalization in regression analysis, *Review of Economics and Statistics*, **52**, 110–113 (1970).

APPENDIX: RIDGE REGRESSION

INTRODUCTION

The regression model is

$$\mathbf{Y} = \mathbf{X}\boldsymbol{\beta} + \mathbf{u} \tag{1}$$

where $\mathbf{Y}$ is an $n \times 1$ vector of observations on a response variable, $\mathbf{X}$ is an $n \times p$ matrix of observations on p explanatory variables, $\boldsymbol{\beta}$ is the $p \times 1$ vector of regression coefficients and $\mathbf{u}$ is an $n \times 1$ vector of residuals satisfying $E(\mathbf{u}) = 0$, $E(\mathbf{u}\mathbf{u}') = \sigma^2 \mathbf{I}$. It is assumed that $\mathbf{X}$ and $\mathbf{Y}$ have been scaled so that $\mathbf{X}'\mathbf{X}$ and $\mathbf{X}'\mathbf{Y}$ are matrices of correlation coefficients.

The least squares estimator for $\boldsymbol{\beta}$ is $\hat{\boldsymbol{\beta}} = (\mathbf{X}'\mathbf{X})^{-1}\mathbf{X}'\mathbf{Y}$. It can be shown that

$$E\left[(\hat{\boldsymbol{\beta}} - \boldsymbol{\beta})'(\hat{\boldsymbol{\beta}} - \boldsymbol{\beta})\right] = \sigma^2 \sum_{j=1}^{p} \lambda_j^{-1}, \tag{2}$$

where $\lambda_1 \geqslant \lambda_2 \geqslant \cdots \geqslant \lambda_p$ are the characteristic roots of $\mathbf{X}'\mathbf{X}$. The left-hand side of Equation (2) is called the total mean square error. It serves as a composite measure of the squared distance of the estimated regression coefficients from their true values.

EFFECT OF MULTICOLLINEARITY

It was argued in Chapter 7 and the appendix to Chapter 7 that multicollinearity is synonymous with small characteristic roots. It follows from Equation (2) that when one or more of the λ's are small, the total mean square error of $\hat{\beta}$ is large suggesting imprecision in the least squares estimation method. The ridge regression approach is an attempt to construct an alternative estimator that has a smaller total mean square error value.

RIDGE REGRESSION ESTIMATORS

Hoerl and Kennard (1970) suggest a class of estimators indexed by a parameter $k > 0$. The estimator is (for a given value of k)

$$\hat{\beta}(k) = (\mathbf{X'X} + k\mathbf{I})^{-1}\mathbf{X'Y} = (\mathbf{X'X} + k\mathbf{I})^{-1}\mathbf{X'X}\hat{\beta}. \tag{3}$$

The expected value of $\hat{\beta}(k)$ is

$$E\left[\hat{\beta}(k)\right] = (\mathbf{X'X} + k\mathbf{I})^{-1}\mathbf{X'X}\beta; \tag{4}$$

the variance–covariance matrix is

$$\text{Var}\left[\hat{\beta}(k)\right] = (\mathbf{X'X} + k\mathbf{I})^{-1}\mathbf{X'X}(\mathbf{X'X} + k\mathbf{I})^{-1}\sigma^2. \tag{5}$$

The residual sum of squares can be written as

$$\left(\mathbf{Y} - \mathbf{X}\hat{\beta}(k)\right)'\left(\mathbf{Y} - \mathbf{X}\hat{\beta}(k)\right) = \left(\mathbf{Y} - \mathbf{X}\hat{\beta}\right)'\left(\mathbf{Y} - \mathbf{X}\hat{\beta}\right)$$

$$+ \left(\hat{\beta}(k) - \hat{\beta}\right)'\mathbf{X'X}\left(\hat{\beta}(k) - \hat{\beta}\right). \tag{6}$$

The total mean square error is

$$E\left[\left(\hat{\beta}(k) - \beta\right)'\left(\hat{\beta}(k) - \beta\right)\right] = \sigma^2 \text{ trace}\left[(\mathbf{X'X} + k\mathbf{I})^{-1}\mathbf{X'X}(\mathbf{X'X} + k\mathbf{I})^{-1}\right]$$

$$+ k^2\beta'(\mathbf{X'X} + k\mathbf{I})^{-2}\beta$$

$$= \sigma^2 \sum_{i=1}^{p} \lambda_i(\lambda_i + k)^{-2} + k^2\beta'(\mathbf{X'X} + k\mathbf{I})^{-2}\beta. \tag{7}$$

Note that the first term on the right-hand side of Equation (7) is the sum of the variances of the components of $\hat{\beta}(k)$ (total variance) and the second term is the square of the bias. For $k > 0$, $\hat{\beta}(k)$ is biased and the bias increases with k. On the other hand, the total variance is a decreasing function of k. The idea of ridge regression is to pick a value of k for which the reduction in total variance is not exceeded by the increase in bias. Hoerl and Kennard (1970), prove that there exists a value of $k > 0$ such that

$$E\left[\left(\hat{\beta}(k) - \beta\right)'\left(\hat{\beta}(k) - \beta\right)\right] < E\left[\left(\hat{\beta} - \beta\right)'\left(\hat{\beta} - \beta\right)\right].$$

They suggest that an appropriate value of k may be selected by observing the ridge trace and some complementary summary statistics for $\hat{\beta}(k)$.

The ridge trace is a simultaneous plot of the components of $\hat{\beta}(k)$ against k, for k in the interval from zero to one. If multicollinearity is a serious problem, the ridge estimators will vary dramatically as k is slowly increased from zero. $\hat{\beta}(k)$ will eventually stabilize. The behavior of $\hat{\beta}(k)$ as a function of k is easily observed from the ridge trace. The value of k selected is the smallest value for which $\hat{\beta}(k)$ is stable. In addition, at the selected value of k, the residual sum of squares should remain close to its minimum value and the variance–covariance matrix of $\hat{\beta}(k)$ should give the appearance of an orthogonal system. (See Hoerl and Kennard, 1970.)

Ridge estimators have been generalized in several ways. They are sometimes generically referred to as shrinkage estimators, because these procedures tend to shrink the estimates of the regression coefficients towards zero. To see one possible generalization, consider the regression model restated in terms of the principal components, **W** (see the Appendix to Chapter 7). The general model takes the form

$$Y = W\alpha + u, \tag{8}$$

where

$$W = XC, \qquad \alpha = C'\beta \tag{9}$$

and

$$C'X'XC = \Lambda, \qquad C'C = CC' = I,$$

Λ is a diagonal matrix consisting of the ordered characteristic roots of $X'X$;

$$\Lambda = \begin{bmatrix} \lambda_1 & 0 & 0...0 & 0 \\ 0 & \lambda_2 & 0...0 & 0 \\ \vdots & & & \\ 0 & 0 & 0...\lambda_{p-1} & 0 \\ 0 & 0 & 0...0 & \lambda_p \end{bmatrix}, \lambda_1 \geqslant \lambda_2 \geqslant \cdots \geqslant \lambda_p.$$

The total mean square error becomes

$$E\left[\left(\hat{\beta}(k)-\beta\right)'\left(\hat{\beta}(k)-\beta\right)\right] = \sigma^2 \sum_{i=1}^{p} \frac{\lambda_i}{(\lambda_i+k)^2} + \sum_{i=1}^{p} \frac{k^2\alpha_i^2}{(\lambda_i+k)^2}, \quad (10)$$

where $\alpha' = (\alpha_1, \alpha_2, \ldots, \alpha_p)$. Instead of taking a single value for k, we can consider several different values k, say $k_1, k_2, \ldots, k_p$. We consider separate ridge parameters (i.e., shrinkage factors) for each of the regression coefficients. The quantity k, instead of being a scalar, is now a vector. The total mean square error given in (10) now becomes

$$E\left[\left(\hat{\beta}(k)-\beta\right)'\left(\hat{\beta}(k)-\beta\right)\right] = \sigma^2 \sum_{1}^{p} \frac{\lambda_i}{(\lambda_i+k_i)^2} + \sum_{1}^{p} \frac{k_i^2\alpha_i^2}{(\lambda_i+k_i)^2}. \quad (11)$$

The total mean square error given in (11) is minimized by taking $k_i = \sigma^2/\alpha_i^2$. An iterative estimation procedure is suggested. At the first step, k_i is computed by using ordinary least squares estimates for σ^2 and α_i. Then, a new value of $\hat{\alpha}(k)$ is computed,

$$\hat{\alpha}(k) = (W'W+k)^{-1}W'Y$$

using $k_1, \ldots, k_p$ from step 1. The process is repeated until successive changes in the components of $\alpha(k)$ are insignificant. Then, using Equation (9), the estimate of β is

$$\hat{\beta}(k) = C\hat{\alpha}(k). \quad (12)$$

The two ridge-type estimators (one value of k, several values of k) defined previously, as well as other related alternatives to ordinary least square estimation, are discussed by Dempster, Schatzoff, and Wermuth (1975). The different estimators are compared and evaluated by Monte Carlo techniques. In general, the choice of the best estimation method for a particular problem depends on the specific model and data. Dempster et al. hint at an analysis that could be used to identify the best estimation method for a given set of data. At the present time, our preference is for the simplest version of the ridge method, a   single ridge parameter k, chosen after an examination of the ridge trace. The appeal of the ridge trace lies in its graphical representation of the effects that multicollinearity has on the estimated coefficients.

CHAPTER 9

Selection of Variables in a Regression Equation

9.1. INTRODUCTION

In our discussion of regression problems so far we have assumed that the variables which go into the equation were chosen in advance. Our analysis involved examining the equation to see whether the functional specification was correct, and whether the assumptions about the error term were valid. The analysis presupposed that the set of variables to be included in the equation had already been decided. In many applications of regression analysis, however, the set of variables to be included in the regression model is not predetermined, and it is often the first part of the analysis to select these variables. There are some occasions when theoretical or other considerations determine the variables to be included in the equation. In those situations the problem of variable selection does not arise. But in situations where there is no clear-cut theory, the problem of selecting variables for a regression equation becomes an important one.

The problems of variable selection and the functional specification of the equation are linked to each other. The questions to be answered while formulating a regression model are, which variables should be included, and in what form should they be included; that is, should they enter the equation as an original variable x, or as some transformed variable such as x^2, $\log x$, or a combination of both? Although ideally both problems should be solved simultaneously, we shall for simplicity propose that they be treated sequentially. We will first determine the variables which will be included in the equation, and after that investigate the exact form in which the variables enter it. This approach is a simplification, but it makes the problem of variable selection more tractable. Once the variables which are to be included in the equation have been selected, we can apply the methods described in the earlier chapters to arrive at the actual form of the equation.

9.2. FORMULATION OF THE PROBLEM

We have q independent variables $X_1, X_2, \ldots, X_q$ and a dependent variable y. A linear model that represents y in terms of q variables is

$$y_i = \beta_0 + \sum_{j=1}^{q} \beta_j x_{ji} + u_i, \qquad (9.1)$$

where β_j are parameters and u_i represents random disturbances. Instead of dealing with the full set of variables (particularly when q is large), we might delete a number of variables and construct an equation with a subset of variables. This chapter is concerned with determining which variables are to be retained in the equation. Let us denote the set of variables retained by $X_1, X_2, \ldots, X_p$ and those deleted by $X_{p+1}, X_{p+2}, \ldots, X_q$. Let us examine the effect of variable deletion under two general conditions:

1. The model which connects y to the x's has all β's $(\beta_0, \beta_1, \ldots, \beta_q)$ nonzero.
2. The model has $\beta_0, \beta_1, \ldots, \beta_p$ nonzero, but $\beta_{p+1}, \beta_{p+2}, \ldots, \beta_q$ zero.

Suppose instead of fitting (9.1) we fit the subset model

$$y_i = \beta_0 + \sum_{1}^{p} \beta_j x_{ji} + u_i. \qquad (9.2)$$

We shall describe the effect of fitting the model to the full and partial set of X's under the two alternative situations described previously. In short, what are the effects of including variables in an equation when they should be properly left out (because the population regression coefficients are zero) and the effect of leaving out variables when they should be included (because the population regression coefficients are not zero)? We will examine the effect of deletion of variables on the estimates of parameters and the predicted values of y. The solution to the problem of variable selection becomes a little clearer once the effects of retaining nonessential variables or the deletion of essential variables in an equation is known.

9.3. CONSEQUENCES OF DELETION OF VARIABLES

Denote the estimates of the regression parameters by $b_0^*, b_1^*, \ldots, b_q^*$ when the model (9.1) is fitted to the full set of variables $X_1, X_2, \ldots, X_q$. Denote the estimates of the regression parameters by $b_0, b_1, \ldots, b_p$ when the model (9.2) is fitted. Let $\hat{y}^*$ and $\hat{y}$ be the predicted values from the full and partial set of variables corresponding to an observation $(x_1, x_2, \ldots, x_q)$. The results can now be summarized as follows (a summary using matrix notation is given in the Appendix): $b_0, b_1, \ldots, b_p$ are biased estimates of $\beta_0, \beta_1, \ldots, \beta_p$ unless the remaining β's in the model ($\beta_{p+1}, \beta_{p+2}, \ldots, \beta_q$) are zero or the variables $(X_1, X_2, \ldots, X_p)$ are orthogonal to the variable set $(X_{p+1}, X_{p+2}, \ldots, X_q)$. The estimates $b_0^*, b_1^*, \ldots, b_p^*$ have less precision than $b_0, b_1, \ldots, b_p$; that is,

$$\text{Var}(b_i^*) \geqslant \text{Var}(b_i), \qquad i = 0, 1, \ldots, p.$$

The variance of the estimates of regression coefficients for variables in the reduced equation are not greater than the variances of the corresponding estimates for the full model. Deletion of variables decreases or, more correctly, never increases, the variances of estimates of the retained regression coefficients. Since b_i are biased and b_i^* are not, a better comparison of the precision of estimates would be obtained by comparing the mean square errors of b_i with the variances of b_i^*. The mean squared errors (MSE) of b_i will be smaller than the variances of b_i^*, only if the deleted variables have regression coefficients smaller in magnitude than the standard deviation of the estimate of the corresponding coefficients. The estimate of σ^2, based on the subset model is generally biased upward.

Let us now look at the effect of deletion of variables on prediction. The prediction $\hat{y}$ is biased unless the deleted variables have zero regression coefficients, or the set of retained variables are orthogonal to the set of deleted variables. The variance of a predicted value from the subset model is smaller than or equal to the variance of the predicted value from the full model; that is,

$$\text{Var}(\hat{y}) \leqslant \text{Var}(\hat{y}^*).$$

The conditions for $\text{MSE}(\hat{y})$ to be smaller than $\text{Var}(\hat{y}^*)$ are identical to the conditions for $\text{MSE}(\hat{b})$ to be smaller than $\text{Var}(\hat{b}^*)$, which we have already stated. For further details refer to Hocking(1974).

The rationale for variable selection can be outlined as follows: Even though the variables deleted have nonzero regression coefficients, the regression coefficients of the retained variables may be estimated with smaller variance from the subset model than from the full model. The same result also holds for the variance of a predicted response. The price paid for deleting variables is in the introduction of bias in the estimates. However, there are conditions (we have described them above), when the MSE of the biased estimates will be smaller than the variance of the unbiased estimates; that is, the gain in precision is not offset by the square of the bias. On the other hand, if some of the retained variables are extraneous or nonessential, that is, have zero coefficients or coefficients whose magnitudes are smaller than the standard deviation of the estimates, then the inclusion of these variables in the equation leads to a loss of precision in estimation and prediction.

9.4. PRELIMINARY REMARKS ON VARIABLE SELECTION

Before discussing actual selection procedures we make two preliminary remarks. First, it is not usually meaningful to speak of the "best set" of variables to be included in the multiple regression equation. There is no unique "best set" of variables. A regression equation can be used for several purposes. The set of variables which may be best for one purpose may not be best for another. The purpose for which a regression equation is constructed should be kept in mind in the variable selection process. We shall show later that the purpose for which an equation is constructed determines the criteria for selecting and evaluating the contributions of different variables.

Second, since there is no best set of variables, there may be several subsets that are adequate and could be used in forming an equation. A good variable selection procedure should point out these several sets rather than generate a so-called single "best" set. The various sets of adequate variables throw light on the structure of data and help us in understanding the underlying process. In fact, the process of variable selection should be viewed as an intensive analysis of the correlational structure of the independent variables and how they individually and jointly affect the dependent variable under study. These two points influence the methodology that we present in connection with variable selection.

9.5. USES OF REGRESSION EQUATIONS

A regression equation has many uses. These may be broadly summarized as follows:

Description and Model Building

A regression equation may be used to describe a given process or as a model for a complex interacting system. The purpose of the equation may be purely descriptive, to clarify the nature of this complex interaction. For this use there are two conflicting requirements: (i) to explain as much of the variation as possible, which points in the direction for inclusion of large numbers of variables; and (ii) to adhere to the principle of parsimony, which suggests that we try for ease of understanding to describe the process with as few variables as possible. In situations where description is the prime goal, we try to choose the smallest number of independent variables that explains the most substantial part of the variation in the dependent variable.

Estimation and Prediction

A regression equation is sometimes constructed for prediction. From the regression equation we want to predict the value of a future observation or estimate the mean response corresponding to a given observation. When a regression equation is used for this purpose, the variables are selected with an eye towards minimizing the MSE of prediction.

Control

A regression equation may be used as a tool for control. The purpose for constructing the equation may be to determine the magnitude by which the value of an independent variable must be altered to obtain a specified value of the dependent variable (target response). Here the regression equation is viewed as a response function, with y as the response variable. For control purposes it is desired that the coefficients of the variables in the equation be measured accurately, that is, the standard errors of the regression coefficients are small.

These are the broad uses of a regression equation. Occasionally these functions overlap and an equation is constructed for some or all of these purposes. The main point to be noted is that the purpose for which the regression equation is constructed determines the criterion that is to be optimized in its formulation. It follows that a subset of variables that may be best for one purpose may not be best for another. The concept of the "best" subset of variables to be included in an equation always requires additional qualification.

9.6.   CRITERIA FOR EVALUATING EQUATIONS

To judge the adequacy of various fitted equations we need a criterion. Several have been proposed in the statistical literature. We describe the two which we consider most useful. An exhaustive list of criteria is found in Hocking (1976).

9.7.   RESIDUAL MEAN SQUARE (RMS)

One measure that is used to judge the adequacy of a fitted equation is the residual mean square (RMS). With a p term equation, the RMS is defined as

$$(RMS)_p = \frac{(SSE)_p}{n-p},$$ (9.3)

where $(SSE)_p$ is the residual sum of squares for a p-term equation. Between two equations, the one with the smaller (RMS) is usually preferred especially if the objective is extrapolation.

It is clear that $(RMS)_p$ is related to the multiple correlation coefficient R_p and the adjusted multiple correlation coefficient R_{ap} which have already been described (Chapter 3) as measures for judging the adequacy of fit of an equation. The relationship between these quantities are given by

$$R_p^2 = 1 - (n-p)\frac{(RMS)_p}{SST}$$ (9.4)

and

$$R_{ap}^2 = 1 - (n-1)\frac{(RMS)_p}{SST},$$ (9.5)

where

$$SST = \sum (y_i - \bar{y})^2.$$

Another measure that is used to judge the adequacy of an equation is called C_p.

9.8. C_p: DEFINITION AND USE

We pointed out earlier that predicted values obtained from a regression equation based on a subset of variables are generally biased. To judge the performance of an equation we should consider the mean square error of the predicted value rather than the variance. J_p measures the standardized total mean squared error of prediction for the observed data and is given by

$$J_p = \frac{1}{\sigma^2}\sum_1^n MSE(\hat{y}_i),$$ (9.6)

where $MSE(\hat{y}_i)$ is the mean squared error of the ith predicted value from a p-term equation, and σ^2 is the variance of the residuals. $MSE(\hat{y}_i)$ has two components, the variance of prediction arising from estimation, and a bias component arising from the deletion of variables. It follows that (Mallows, 1973) J_p is the sum of two components V_p and B_p/σ^2, where V_p and B_p are the "variance" and "bias" components, respectively. J_p can be expressed as

$$J_p = \frac{E(SSE)_p}{\sigma^2} + (2p - n).$$

(9.7)

To estimate J_p we use C_p defined as

$$C_p = \frac{(SSE)_p}{\hat{\sigma}^2} + (2p - n),$$

(9.8)

where $\hat{\sigma}^2$ is an estimate of σ^2, and is usually obtained from the linear model with the full set of q variables. It can be shown that the expected value of C_p is p, when there is no bias in the fitted equation using p variables. Consequently, the deviation of C_p from p can be used as a measure of bias. The C_p statistic, therefore, measures the performance of the variables in terms of the standardized mean square error of prediction. It takes into account both the bias as well as the variance. Subsets of variables that produce values of C_p that are close to p are the desirable subsets. The selection of "good" subsets is done graphically. For the various subsets a graph of C_p is plotted against p. The line $C_p = p$ is also drawn on the graph. Sets of variables corresponding to points close to the line $C_p = p$ are the good or desirable subsets of variables to form an equation. The use of C_p plots is illustrated and discussed in more detail in the example that is found in section 9.16. A very thorough treatment of the C_p statistic is given in Daniel and Wood (1971).

9.9.   EXAMINATION OF COLLINEARITY

In discussing variable selection procedures we distinguish between two broad situations. These are:

1.   The independent variables are not collinear; that is, no strong evidence of multicollinearity.

2.   The independent variables are collinear; that is, the data is highly multicollinear.Depending on the correlation structure of the independent variables we propose different approaches to the variable selection procedure. If the data analyzed is not collinear we proceed in one manner, and if collinear we proceed in another. As a first step in variable selection procedure we recommend calculating the characteristic roots of the correlation matrix of the independent variables. As has already been explained in Chapter 7, the presence of small characteristic roots indicates collinear-

ity. Besides the individual roots we also look at the sum of the reciprocals of the characteristic roots. If any of the individual characteristic roots are less than 0.01, or the sum of the reciprocals of the roots is greater than, say five times the number of explanatory variables in the problem, then we say that the variables are collinear. If the above conditions do not hold, the variables are regarded as noncollinear.

9.10. EVALUATING ALL POSSIBLE EQUATIONS

The first procedure described is very direct and applies equally well to both collinear and noncollinear data. The procedure involves fitting all possible subset equations to a given body of data. With q variables the total number of equations fitted is 2^q. Included are an equation that contains all the variables, and another that contains no variables. The latter is simply $\hat{y}_i = \bar{y}$. This method clearly gives an analyst the maximum amount of information available concerning the nature of relationships between y and the set of x's. However, the number of equations and supplementary information that must be looked at may be prohibitive. Even with only six explanatory variables, there are 64 (2^6) equations to consider; with seven variables the number grows to 128 (2^7), neither feasible nor practical.

When using the method of all subset regressions, the most promising are isolated using either C_p or RMS. These regressions are then carefully analyzed by examining the residuals for outliers, autocorrelation, or the need for transformations before deciding on the final subset of variables to retain. The various subsets that are investigated may suggest interpretations of the data that might have been overlooked in a more restricted variable selection approach.

When the number of variables is large, the evaluation of all possible equations may not be practically feasible. Certain shortcuts have been suggested (Furnival and Wilson, 1974; La Motte and Hocking, 1970) which do not involve computing the entire set of equations while searching for the desirable subsets. But with a large number of variables these methods still involve a considerable amount of computation. There are variable selection procedures which do not require the evaluation of all possible equations. Employing these procedures will not provide the analyst with as much information as the fitting of all possible equations, but it will entail considerably less computation, and may be the only available practical solution. The procedures are generically called stepwise procedures. These procedures are quite efficient with noncollinear data. We do not, however, recommend the use of stepwise procedures with collinear data.

9.11.　SELECTION OF VARIABLES: STEPWISE PROCEDURE

For cases when there are a large number of potential explanatory variables, a set of procedures that does not involve computing of all possible equations has been proposed. These procedures have the feature that the variables are introduced or deleted from the equation one at a time, and involves examining only a subset of all possible equations. With q variables these procedures will involve evaluation of at most $(q+1)$ equations as contrasted with the evaluation of 2^q equations necessary for examining all possible equations. The procedures can be classified into two broad categories (i) the forward selection procedure (FS) and (ii) the backward elimination procedure (BE). There is also a very popular modification of the FS procedure called the stepwise method. The three procedures are described and compared in the following sections.

9.12.　FORWARD SELECTION PROCEDURE

This procedure starts with an equation containing no explanatory variables, only a constant term. The first variable included in the equation is the one which has the highest simple correlation with the dependent variable y. If the regression coefficient of this variable is significantly different from zero it is retained in the equation, and a search for a second variable is made. The variable that enters the equation as the second variable is one which has the highest correlation with y, after y has been adjusted for the effect of the first variable, that is, the variable with the highest simple correlation coefficient with the residuals from step 1. The significance of the regression coefficient of the second variable is then tested. If the regression coefficient is significant, a search for a third variable is made in the same way. The procedure is terminated when the last variable entering the equation has insignificant regression coefficient or all the variables are included in the equation. The significance of the regression coefficient of the last variable introduced in the equation is judged by the standard t statistic computed from the latest equation. Most forward selection algorithms codes use a low t cutoff value for testing the coefficient of the newly entered variable; consequently, the forward selection procedure goes through the full set of variables and provides us with q possible equations.

9.13.　BACKWARD ELIMINATION PROCEDURE

This procedure starts with the full equation and successively drops one variable at a time. The variables are dropped on the basis of their contribution to the reduction of error sum of squares. The first variable deleted is the one with the smallest contribution to the reduction of error sum of squares. This is equivalent to deleting the variable which has the smallest t ratio (the ratio of the regression coefficient to the standard error of the coefficient) in the equation. If all the t ratios are significant, the full set of variables is retained in the equation. Assuming that there are one or more variables that have insignificant t ratios, the procedure operates by dropping the variable with the smallest insignificant t ratio. The equation with the remaining $(q-1)$ variables is then fitted and the t ratios for the new regression coefficients are examined. The procedure is terminated when all the t ratios are significant or all but one variable has been deleted. In most backward elimination algorithms the cutoff value for the t ratio is set so high that the procedure runs through the whole set of variables, that is, starting with the q-variable equation and ending up with a one-variable equation. The backward elimination procedure involves fitting at most q regression equations.

9.14.   STEPWISE METHOD

The stepwise method is essentially a forward selection procedure, but with the added proviso that at each stage the possibility of deleting a variable, as in backward elimination, is considered. In this procedure a variable that entered in the earlier stages of selection may be eliminated at later stages. The calculations made for inclusion and deletion of variables are the same as FS and BE procedures. Often different levels of significance are assumed for inclusion and exclusion of variables from the equation.

9.15.   GENERAL COMMENTS ON STEPWISE
PROCEDURES

Stepwise procedures for the selection of variables in a regression problem should be used with caution. These procedures should not be used mechanically to determine the "best" variables. The order in which the variables enter or leave the equation in stepwise procedures should not be interpreted as reflecting the relative importance of the variables. If these caveats are kept in mind, the stepwise procedures are useful tools for variable selection in noncollinear situations. All three procedures will give nearly the same selection of variables with noncollinear data. They entail much less computing than the analysis of all possible equations.

Several stopping rules have been proposed for stepwise procedures. A stopping rule that has been reported to be quite effective is as follows:

In FS: Stop if minimum t ratio is less than 1.
In BE: Stop if minimum t ratio is greater than 1.

In the following example we illustrate the effect of different stopping rules in variable selection.

We recommend the BE procedure over FS procedure for variable selection. One obvious reason is that in BE procedure the equation with the full variable set is calculated and available for inspection even though it may not be used as the final equation. Although we do not recommend the use of stepwise procedures in a collinear situation, the BE procedure is better able to handle multicollinearity than the FS procedure (Mantel, 1970).

In an application of stepwise procedures several equations are generated, each equation containing a different number of variables. The various equations generated can then be evaluated using a statistic such as C_p or RMS. The residuals for the various equations should also be examined. Equations with unsatisfactory residual plots are rejected. Only a total and comprehensive analysis will provide an adequate selection of variables and a useful regression equation. This approach to variable selection is illustrated in the following example.

9.16.   A STUDY OF SUPERVISOR PERFORMANCE

To illustrate variable selection procedures in a noncollinear situation, consider the example discussed earlier in Chapter 3. A regression equation was needed to study the qualities which led to the characterization of good supervisors by the people being supervised. The equation is to be constructed in an attempt to understand the supervising process and the relative importance of the different variables. In terms of the use for the regression equation this would imply that we want accurate estimates of the regression coefficients in contrast to an equation which is to be used only for prediction. The variables in the problem are:

Y Overall rating of job being done by supervisor
X_1 Handles employee complaints
X_2 Does not allow special privileges
X_3 Opportunity to learn new things
X_4 Raises based on performance
X_5 Too critical of poor performances
X_6 Rate of advancement to better jobs

The correlation matrix for the data is given in Table 9.1. As a first step we calculate the characteristic roots of the correlation matrix. The characteris-

Table 9.1.   *Correlation matrix for the data of Table 3.2*

	X1	X2	X3	X4	X5	X6
X1	1.000					
X2	0.558	1.000				
X3	0.597	0.493	1.000			
X4	0.669	0.445	0.640	1.000		
X5	0.188	0.147	0.116	0.377	1.000	
X6	0.225	0.343	0.532	0.574	0.283	1.000

tic roots are

$$\lambda_1 = 3.236, \quad \lambda_2 = 1.033, \quad \lambda_3 = 0.750,$$
$$\lambda_4 = 0.535, \quad \lambda_5 = 0.253, \quad \lambda_6 = 0.193.$$

The sum of the reciprocals of the characteristic roots is 13.624. Since none of the roots are small, and the sum of the reciprocals of the roots is only twice the number of variables, we conclude that the data in the present example is not seriously collinear and we can apply the stepwise procedures just described.

The result of forward selection procedure is given in Table 9.2. For successive equations we show the variables present, the RMS and the value of C_p statistic. The last column shows the rank of the subset obtained by FS relative to best subset (on the basis of RMS) of same size. The value of p is the number of explanatory variables in the equation including a constant term. Two stopping rules are used. These are

1. Stop if minimum absolute t ratio is less than $t_{.05}(n-p)$.
2. Stop if minimum absolute t ratio is less than 1.

The first rule is more stringent and terminates with variables X_1 and X_3. The second rule is less stringent and terminates with variables X_1, X_3 and X_6.

The results of applying BE procedure is presented in Table 9.3, which is identical in structure to Table 9.2.

Table 9.2.   *Variables selected by the forward selection method*

Variables in equation	p	RMS	C_p	Rank
X_1	2	6.993	1.41	1
$X_1 X_3$	3	6.817	1.11	1
$X_1 X_3 X_6$	4	6.734	1.60	1
$X_1 X_3 X_6 X_2$	5	6.820	3.28	1
$X_1 X_3 X_6 X_2 X_4$	6	6.928	5.07	1
$X_1 X_3 X_6 X_2 X_4 X_5$	7	7.068	7.00	—

Table 9.3.   Variables selected by backward elimination method

Variables in equation	p	RMS	C_p	Rank
$X_1X_2X_3X_4X_5X_6$	7	7.068	7.00	—
$X_1X_2X_3X_4X_6$	6	6.928	5.07	1
$X_1X_2X_3X_6$	5	6.820	3.28	1
$X_1X_3X_6$	4	6.734	1.60	1
X_1X_3	3	6.817	1.11	1
X_1	2	6.993	1.41	1

For BE we will use the stopping rules.

1. Stop if minimum absolute t ratio is greater than $t_{.05}(n-p)$.
2. Stop if minimum absolute t ratio is greater than 1.

With the first stopping rule the variables selected are X_1 and X_3. With the second stopping rule the variables selected are X_1, X_3, and X_6. The FS and BE give identical equations for this problem, but this is not always the case. An application of the stepwise procedure yielded the same results. To describe the supervisor performance, we therefore recommend the equation

$$Y = 13.58 + 0.62X_1 + 0.31X_3 - 0.19X_6.$$

The residual plots (not shown) for the preceding equation are satisfactory.

Since the present problem has only six variables, the total number of equations that can be fitted which contain at least one variable is 63. The C_p values for all 63 equations are shown in Table 9.4. The C_p values are plotted against p in Figure 9.1. The best subsets of variables based on C_p values are given in Table 9.5.

It is seen that the subsets selected by C_p are different from those arrived at by the stepwise procedures as well as those selected on the basis of residual mean square. This anomaly suggests up an important point concerning the C_p statistic which the reader should bear in mind. For applications of the C_p statistic, an estimate of σ^2 is required. Usually the estimate of σ^2 is obtained from the residual sum of squares from the full model. If the full model has a large number of variables with no explanatory power (i.e., population regression coefficients are zero) the estimate of σ^2 from the residual sum of squares for the full model would be large. The loss in degrees of freedom for the divisor would not be balanced by a reduction in the error sum of squares. If $\hat{\sigma}^2$ is large, then the value of C_p is small. For C_p to work properly, a good estimate of σ^2 must be available. When a good estimate of σ^2 is not available, C_p is of only limited usefulness. In our

Table 9.4. Values of C_p statistic (all possible equations)

Variables	C_p	Variables	C_p	Variables	C_p	Variables	C_p
1	1.41	1 5	3.41	1 6	3.33	1 5 6	5.32
2	44.40	2 5	45.62	2 6	46.39	2 5 6	47.91
1 2	3.26	1 2 5	5.26	1 2 6	5.22	1 2 5 6	7.22
3	26.56	3 5	27.94	3 6	24.82	3 5 6	25.02
1 3	1.11	1 3 5	3.11	1 3 6	1.60	1 3 5 6	3.46
2 3	26.96	2 3 5	28.53	2 3 6	24.62	2 3 5 6	25.11
1 2 3	2.51	1 2 3 5	4.51	1 2 3 6	3.28	1 2 3 5 6	5.14
4	30.06	4 5	31.62	4 6	27.73	4 5 6	29.50
1 4	3.19	1 4 5	5.16	1 4 6	4.70	1 4 5 6	6.69
2 4	29.20	2 4 5	30.82	2 4 6	25.91	2 4 5 6	27.74
1 2 4	4.99	1 2 4 5	6.97	1 2 4 6	6.63	1 2 4 5 6	8.61
3 4	23.25	3 4 5	25.23	3 4 6	16.50	3 4 5 6	18.42
1 3 4	3.09	1 3 4 5	5.09	1 3 4 6	3.35	1 3 4 5 6	5.29
2 3 4	24.56	2 3 4 5	26.53	2 3 4 6	17.57	2 3 4 5 6	19.51
1 2 3 4	4.49	1 2 3 4 5	6.48	1 2 3 4 6	5.07	1 2 3 4 5 6	7.00
5	57.91	6	57.95	5 6	58.76		

present example, the RMS for the full model with six variables is larger than the RMS for the model with three variables X_1, X_3, X_6. Consequently, the C_p values are distorted, and not very useful in variable selection in the present case. The type of situation we have described can be spotted by looking at the RMS for different values of p. RMS will at first tend to decrease with p, but increase at later stages. This behavior indicates that the latter variables are not contributing significantly to the reduction of error sum of squares. Useful application of C_p requires a parallel monitoring of RMS to avoid distortions.

9.17. VARIABLE SELECTION WITH COLLINEAR DATA

In Chapter 7 it was pointed out that serious distortions are introduced in standard analysis with collinear data. Consequently, we recommend a different set of procedures for selecting variables in these situations. Collinearity is indicated when the correlation matrix has one or more small characteristic roots. With a small number of collinear variables we can evaluate all possible equations and select an equation by methods that have already been described. But with a larger number of variables this method is not feasible.

Two different approaches have been proposed to the problem. The first approach, which we do not recommend, tries to break down the collinearity of the data by deleting variables. The collinear structure present in the

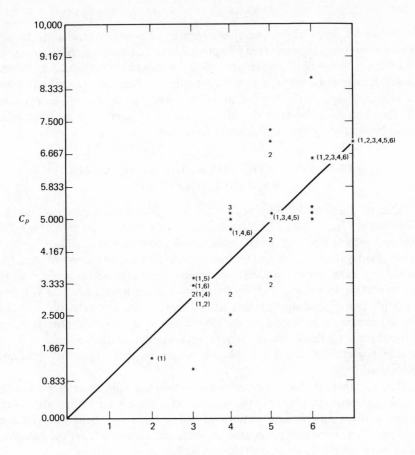

Fig. 9.1. C_p plot.

Table 9.5. **Variables selected on the basis of C_p statistic**

Variables in equation	p	RMS	C_p	Rank
X_1	2	6.993	1.41	1
X_1X_4	3	7.093	3.19	2
$X_1X_4X_6$	4	7.163	4.70	5
$X_1X_3X_4X_5$	5	7.080	5.09	6
$X_1X_2X_3X_4X_5$	6	7.139	6.48	4
$X_1X_2X_3X_4X_5X_6$	7	7.068	7.00	—

207

variables is revealed by the characteristic vectors corresponding to very small characteristic roots (see Chapters 7 and 8). Once the collinearities are identified, a set of variables can then be deleted to produce a reduced noncollinear data set. We can then apply the methods described earlier. The second approach, one which we do recommend, uses ridge regression as the main tool. We assume that the reader is familiar with the basic terms and concepts of ridge regression (Chapter 8).

9.18. APPLICATION OF RIDGE REGRESSION TO VARIABLE SELECTION

One of the goals of ridge regression is to produce a regression equation with stable coefficients. The coefficients are stable in the sense that they are not affected by slight variations in the estimation data. The objectives of a good variable selection procedure are (i) to select a set of variables which provides a clear understanding of the process under study, and (ii) to formulate an equation which provides accurate forecasts of the response variable corresponding to values of the explanatory variables not included in the study. It is seen that the objectives of a good variable selection procedure and ridge regression are very similar, and consequently, one (ridge regression) can be employed to accomplish the other (variable selection).

The variable selection is done by examining the ridge trace, a plot of the ridge regression coefficients against the ridge parameter k. For a collinear system, the characteristic pattern of ridge trace has been described in Chapter 8. The ridge trace is used to eliminate variables from the equation. The rules for elimination are (Hoerl and Kennard, 1970):

1. Eliminate variables whose coefficients are stable but small. Since ridge regression is applied to standardized data, the magnitude of the various coefficients are directly comparable.
2. Eliminate variables with unstable coefficients that do not hold their predicting power, that is, unstable coefficients that tend to zero.
3. Eliminate one or more variables with unstable coefficients. The variables remaining from the original set, say p in number, are used to form the regression equation.

The subset of variables remaining after elimination should be examined to see if they constitute an orthogonal set. A graphical method for doing this is obtained by plotting $\sum_{i}^{p} \hat{\beta}_i^2(k)$ ($\hat{\beta}_i(k)$ is the ridge coefficient for the ith variable corresponding to the ridge parameter k) against k. Geometrically, $\sum_i \hat{\beta}_i^2(k)$ represents the squared distance of the ridge estimates from the origin. For an orthogonal system, the squared distance of the ridge coefficients from the origin should be $\left(\sum_i \hat{\beta}_i^2(0)\right)/(1+k)^2$, where $\hat{\beta}_i(0)$ is the ordinary least square estimate, and k the ridge parameter. If the retained variables constitute approximately an orthogonal set, then the graph of $\sum_i \hat{\beta}_i^2(k)$ and $\sum_i \hat{\beta}_i^2(0))/(1+k)^2$ plotted against k should be nearly identical.

If the graphs are identical we proceed to the next step in the analysis, but if they are dissimilar additional variables have to be deleted. The variables eliminated are those with unstable coefficients. This procedure should produce an approximately orthogonal system. The final equation is then estimated by the ridge method from the full model rather than fitting a least square equation to the reduced model. This is equivalent to discarding variables by setting the values of all discarded variables at their sample means for all predictions. We illustrate this procedure by an example.

9.19.   SELECTION OF VARIABLES IN AN AIR POLLUTION STUDY

McDonald and Schwing (1973) present a study which relates total mortality to climate, socioeconomic, and pollution variables. Fifteen independent variables selected for the study are listed in Table 9.6. The dependent variable is the total age adjusted mortality from all causes.

We will not comment on the epidemiological aspects of the study, but merely use the data as an illustrative example for variable selection. A very detailed discussion of the proplem is presented by McDonald and Schwing in their paper and we refer the interested reader to it for more information.

Table 9.7 is the correlation matrix of the independent variables. As can be expected from the nature of the variables, some of them are highly correlated with each other. The evidence of collinearity is clearly seen if we examine the characteristic roots of the correlation matrix. The characteristic roots are

$$\begin{array}{lll} \lambda_1 = 4.5272, & \lambda_6 = .9605, & \lambda_{11} = .1665, \\ \lambda_2 = 2.7547, & \lambda_7 = .6124, & \lambda_{12} = .1275, \\ \lambda_3 = 2.0545, & \lambda_8 = .4729, & \lambda_{13} = .1142, \\ \lambda_4 = 1.3487, & \lambda_9 = .3708, & \lambda_{14} = .0460, \\ \lambda_5 = 1.2227, & \lambda_{10} = .2163, & \lambda_{15} = .0049. \end{array}$$

Table 9.6. *Description of variables, means, and standard deviations (n = 60)*

Variable number	Description	Mean	SD
1	Mean annual precipitation in inches	37.37	9.98
2	Mean January temperature in Fahrenheit	33.98	10.17
3	Mean July temperature in Fahrenheit	74.58	4.76
4	Percent of population over 65 years of age	8.80	1.46
5	Population per household	3.26	0.14
6	Median school years completed	10.97	0.85
7	Percent of housing units which are sound	80.92	5.15
8	Population per square mile	3876.05	1454.10
9	Percent of nonwhite population	11.87	8.92
10	Percent employment in white collar jobs	46.08	4.61
11	Percent of families with income under $3000	14.37	4.16
12	Relative pollution potential of hydrocarbons	37.85	91.98
13	Relative pollution potential of oxides of nitrogen	22.65	46.33
14	Relative pollution potential of sulfur dioxide	53.77	63.39
15	Percent relative humidity	57.67	5.37

Source: McDonald and Schwing (1973).

There are two very small roots; the largest root is nearly 1000 times larger than the smallest root. The sum of the reciprocals of the roots is 263.06, which is nearly 17 times the number of variables. The data shows strong evidence of collinearity. Examination of the characteristic vector corresponding to the smallest characteristic root shows a strong linear relationship between variables 12, 13, 14. In fact, the relationship connecting them is

$$-.689x_{12} + .712x_{13} - .108x_{14} \doteq 0,$$

where x_i $(i = 1,2,\ldots,15)$ is the standardized variable corresponding to X_i $(i = 1,2,\ldots,15)$ and the coefficients of all other variables are approximately zero. This is obtained from the characteristic vector corresponding to the root λ_{15} $(= .0049)$. (See Chapter 7.)

We now eliminate variables on the basis of ridge trace (Figure 9.2). Following the first criterion we eliminate variables 4, 7, 10, 11, and 15. These variables all have fairly stable coefficients, as shown by the flatness of their ridge traces, but are very small. The second criterion suggests eliminating variables with unstable coefficients which tend to zero. Examination of the ridge trace reveals that variables 12 and 13 fall in this category. Third criterion suggests eliminating variables with unstable coef-

Table 9.7. Correlation matrix for the variables in Table 9.6

	2	3	4	5	6	7	8	9	10	11	12	13	14	15	16
1	.0922	.5033	.1011	.2634	-.4904	-.4903	-.0035	.4132	-.2973	.5066	-.5318	-.4873	-.1069	-.0773	.5095
2		.3463	-.3981	-.2092	.1163	.0139	-.1001	.4538	.2380	.5653	.3508	.3210	-.1078	.0679	-.0300
3			-.4340	.2623	-.2385	-.4155	-.0610	.5753	-.0214	.6193	-.3565	-.3377	-.0993	-.4528	.2770
4				-.5091	-.1389	.0649	.1620	-.6378	-.1177	-.3098	-.0205	-.0021	.0172	.1124	-.1746
5					-.3951	-.4095	-.1843	.4194	-.4257	.2599	-.3882	-.3584	-.0041	-.1357	.3573
6						.5515	-.2439	-.2088	.7032	-.4033	.2868	.2244	-.2343	.1765	-.5110
7							.1806	-.4091	.3376	-.6806	.3859	.3476	.1180	.1224	-.4248
8								-.0057	-.0318	-.1629	.1203	.1653	.4321	-.1250	.2655
9									-.0044	.7049	-.0259	.0184	.1593	-.1180	.6437
10										-.1852	.2037	.1600	-.0685	.0607	-.2848
11											-.1298	-.1025	-.0965	-.1522	.4105
12												.9838	.2823	-.0202	-.1772
13													.4094	-.0459	-.0774
14														-.1026	.4259
15															-.0885

Source: McDonald and Schwing (1973).

211

RIDGE TRACE: TOTAL MORTALITY (Variables 1–15)

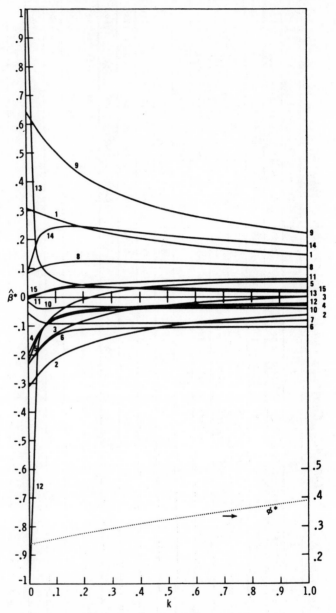

Fig. 9.2. Ridge trace: Total mortality (variables 1–15).
Source: McDonald and Schwing (1973).

ficients, and from the trace it is seen that variables 3 and 5 fall in this category. The variables remaining after the screening by ridge trace are $(1, 2, 6, 8, 9, 14)$.

To see if the variables retained can be regarded as an orthogonal set we compare the observed squared length of the coefficient vector to the expected squared length of the coefficient vector which would result from an orthogonal set. The plot of $\sum_i \hat{\beta}_i^2(k)$ and $\left(\sum_i \hat{\beta}_i^2(0) \right) / (1+k)^2$ against k are both shown in Figure 9.3, the first by continuous and the second by dotted lines. The two graphs seem identical showing that the retained variables constitute an approximately orthogonal set.

Examination of Figure 9.2 shows that the ridge trace is stable at $k = 0.2$. The fitted equation in terms of the standardized variables is

$$y = .243 x_1 - .168 x_2 - .114 x_6 + .123 x_8 + .423 x_9 + .243 x_{14}.$$

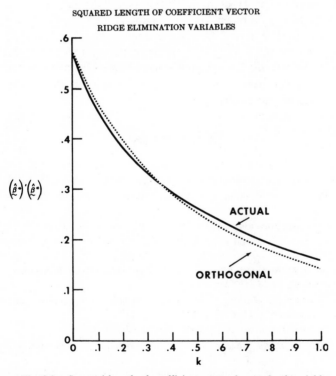

SQUARED LENGTH OF COEFFICIENT VECTOR

RIDGE ELIMINATION VARIABLES

Fig. 9.3. Squared length of coefficient vector for retained variables.
Source: McDonald and Schwing (1973).

From the set of 15 variables we have chosen a set of six variables to form an equation with stable coefficients. The values of the statistics that have been proposed for judging the equation are $C_p = 5.52$ and $R^2 = 0.571$. The fitted equation is seen to perform quite adequately.

We hope it is clear from our discussion that variable selection is a mixture of art and science, and should be performed with care and caution. We have outlined a set of approaches rather than prescribing a formal procedure. In conclusion, we must emphasize the point made earlier that variable selection should not be performed mechanically as an end in itself, but rather as an exploration into the structure of the data analyzed, and as in all true explorations, the explorer is guided by theory, intuition, and common sense.

BIBLIOGRAPHIC NOTES

There is a vast amount of literature on variable selection scattered in statistical journals. A very comprehensive review with an extensive bibliography is to be found in Hocking (1976). A detailed treatment on variable selection with special emphasis on C_p statistic is found in Chapter 6 of the text by Daniel and Wood (1971). Refinements on the application of C_p statistic is given by Mallows (1973). Stepwise procedures are discussed in Chapter 6 in the text by Draper and Smith (1966). Use of ridge regression in connection with variable selection is discussed by Hoerl and Kennard (1970), McDonald and Schwing (1973).

REFERENCES

Daniel, C. and F. S. Wood, *Fitting Equations to Data,* Wiley, New York, 1971.

Draper, N. R. and H. Smith, *Applied Regression Analysis,* Wiley, New York, 1966.

Furnival, G. M. and R. W. Wilson, Jr., Regression by leaps and bounds, *Technometrics,* **16,** 499–512 (1974).

Hocking, R. R., Misspecification in regression, *The American Statistician,* **28,** 39–40 (1974).

Hocking R. R., The analysis and selection of variables in linear regression, *Biometrics,* **32,** 1–49 (1976).

Hoerl, A. E. and R. W. Kennard, Ridge regression: Applications to nonorthogonal problems, *Technometrics,* **12,** 69–82 (1970).

La Motte, L. R. and R. R. Hocking, Computational efficiency in the selection of regression variables, *Technometrics,* **12,** 83–93 (1970).

Mallows, C. L., Some comments on C_p, *Technometrics,* **15,** 661–75 (1973).

Mantel, N., Why stepdown procedures in variable selection, *Technometrics,* **12,** 591–612 (1970).

McDonald, G. C. and R. C. Schwing, Instabilities of regression estimates relating air pollution to mortality, *Technometrics,* **15,** 463–81 (1973).

APPENDIX

The following discussion notes the effects of an incorrect model specification on the estimates of the regression coefficients and predicted values.
Define the following matrix and vectors:

$$
\mathbf{X} = \begin{bmatrix}
x_{01} & x_{11} & \cdots & x_{p1} & \vdots & \cdots & x_{q1} \\
x_{02} & x_{12} & \cdots & x_{p2} & \vdots & \cdots & x_{q2} \\
\cdot & & & & \vdots & & \\
\cdot & & & & \vdots & & \\
\cdot & & & & \vdots & & \\
x_{0n} & x_{1n} & \cdots & x_{pn} & \vdots & \cdots & x_{qn}
\end{bmatrix}, \quad
\mathbf{Y} = \begin{bmatrix} y_1 \\ \cdot \\ \cdot \\ \cdot \\ y_n \end{bmatrix},
$$

$$
\boldsymbol{\beta} = \begin{bmatrix}
\beta_0 \\ \beta_1 \\ \cdot \\ \cdot \\ \beta_p \\ \hline \beta_{p+1} \\ \cdot \\ \cdot \\ \cdot \\ \beta_q
\end{bmatrix}, \quad
\mathbf{u} = \begin{bmatrix} u_1 \\ u_2 \\ \cdot \\ \cdot \\ \cdot \\ u_n \end{bmatrix},
$$

where $x_{0i} = 1$ for all i $(i = 1, \ldots, n)$.

The matrix $\mathbf{X}$ which has n rows and $(q + 1)$ columns is partitioned into two submatrices $\mathbf{X}_p$ and $\mathbf{X}_r$, of dimensions $(n \times (p + 1))$ and $(n \times r)$, where $r = q - p$. The vector $\boldsymbol{\beta}$ is similarly partitioned into $\boldsymbol{\beta}_p$, $\boldsymbol{\beta}_r$ which has $(p + 1)$ and r components, respectively.

The full linear model containing all q variables is given by

$$
\mathbf{Y} = \mathbf{X}\boldsymbol{\beta} + \mathbf{u} = \mathbf{X}_p\boldsymbol{\beta}_p + \mathbf{X}_r\boldsymbol{\beta}_r + \mathbf{u}, \tag{1}
$$

where u_i's are residuals which are independently normally distributed with zero means and unit variance.

The linear model containing only p variables (i.e., an equation with $(p + 1)$ terms) is

$$
\mathbf{Y} = \mathbf{X}_p\boldsymbol{\beta}_p + \mathbf{u}. \tag{2}
$$

Let us denote the least squares estimate of $\boldsymbol{\beta}$ obtained from the full model

(1) by b^* where

$$b^* = \begin{pmatrix} b_p^* \\ b_r^* \end{pmatrix} = (X'X)^{-1}X'Y.$$

The estimate $\mathbf{b}_p$ of $\boldsymbol{\beta}_p$ obtained from the subset model is given by

$$\mathbf{b}_p = (X_p'X_p)^{-1}X_p'Y.$$

Let $\hat{\sigma}_*^2$ and $\hat{\sigma}_p^2$ denote the estimates of σ^2 obtained from (1) and (2), respectively. Then it follows

$$\hat{\sigma}_*^2 = \frac{Y'Y - b^{*\prime}X'Y}{n - q - 1}$$

and

$$\hat{\sigma}_p^2 = \frac{Y'Y - b_p'X_p'Y}{n - p - 1}$$

It is known from standard theory that b^* and $\hat{\sigma}_*^2$ are unbiased estimates of β and σ^2. It can be shown that

$$E(\mathbf{b}_p) = \boldsymbol{\beta}_p + A\boldsymbol{\beta}_r,$$

where

$$A = (X_p'X_p)^{-1}X_p'X_r.$$

Further

$$\mathrm{Var}(\mathbf{b}_p) = (X_p'X_p)^{-1}\sigma^2,$$

$$\mathrm{Var}(\mathbf{b}^*) = (X'X)^{-1}\sigma^2,$$

and

$$\mathrm{MSE}(\mathbf{b}_p) = (X_p'X_p)^{-1}\sigma^2 + A\boldsymbol{\beta}_r\boldsymbol{\beta}_r'A'.$$

We can summarize the properties of $\mathbf{b}_p$ and $\mathbf{b}_p^*$ as follows (Hocking, 1976):

1. $\mathbf{b}_p$ is a biased estimate of $\boldsymbol{\beta}_p$ unless (i) $\boldsymbol{\beta}_r = 0$ or (ii) $X_p'X_r = 0$.
2. The matrix $\mathrm{Var}(\mathbf{b}_p^*) - \mathrm{Var}(\mathbf{b}_p)$ is positive semidefinite; that is, vari-

ances of the least squares estimates of regression coefficients obtained from the full model are larger than the corresponding variances of the estimates obtained from the subset model. In other words, the deletion of variables always results in smaller variances for the estimates of the regression coefficients of the remaining variables.

3. If the matrix $\text{Var}(\mathbf{b}_r^*) - \boldsymbol{\beta}_r\boldsymbol{\beta}_r'$ is positive semidefinite, then the matrix $\text{Var}(\mathbf{b}_p^*) - \text{MSE}(\mathbf{b}_p)$ is positive semidefinite. This means that the least square estimates of regression coefficients obtained from the subset model have smaller mean squares than estimates obtained from the full model, when the variables deleted have regression coefficients which are smaller than the standard deviation of the estimates of the coefficients.

4. $\hat{\sigma}_p^2$ is generally biased upward as an estimate of σ^2.

To see the effect of model misspecification on prediction, let us examine the prediction corresponding to an observation, say $\mathbf{x}' = (\mathbf{x}_p'\mathbf{x}_r')$. Let $\hat{y}^*$ denote the predicted value corresponding to $\mathbf{x}'$ when the full set of variables are used. Then $\hat{y}^* = \mathbf{x}'\mathbf{b}^*$ with mean $\mathbf{x}'\boldsymbol{\beta}$ and prediction variance $\text{Var}(\hat{y}^*)$:

$$\text{Var}(\hat{y}^*) = \sigma^2\left(1 + \mathbf{x}'(\mathbf{X}'\mathbf{X})^{-1}\mathbf{x}\right).$$

On the other hand, if the subset model (2) is used, the estimated predicted value $\hat{y} = \mathbf{x}_p'\mathbf{b}_p$ with mean $E(\hat{y})$:

$$E(\hat{y}) = \mathbf{x}_p'\boldsymbol{\beta}_p + \mathbf{x}_p'\mathbf{A}\boldsymbol{\beta}_r$$

and prediction variance

$$\text{Var}(\hat{y}) = \sigma^2\left(1 + \mathbf{x}_p'(\mathbf{X}_p'\mathbf{X}_p)^{-1}\mathbf{x}_p\right).$$

The prediction mean squared error is given by

$$\text{MSE}(\hat{y}) = \sigma^2\left(1 + \mathbf{x}_p'(\mathbf{X}_p'\mathbf{X}_p)^{-1}\mathbf{x}_p\right) + \left(\mathbf{x}_p'\mathbf{A}\boldsymbol{\beta}_r - \mathbf{x}_r'\boldsymbol{\beta}_r\right)^2.$$

The properties of $\hat{y}^*$ and $\hat{y}$ can be summarized as follows:

5. $\hat{y}$ is biased unless $\mathbf{X}_p'\mathbf{X}_r\boldsymbol{\beta}_r = 0$.
6. $\text{Var}(\hat{y}^*) \geqslant \text{Var}(\hat{y})$.
7. If the matrix $\text{Var}(\mathbf{b}_r^*) - \boldsymbol{\beta}_r\boldsymbol{\beta}_r'$ is positive semidefinite, then $\text{Var}(\hat{y}^*) \geqslant \text{MSE}(\hat{y})$.

The significance and interpretation of these results in the context of variable selection are given in the main body of the accompanying chapter.

Statistical Tables

Table A.1.   Values P_x and x on the normal distribution

$N(0, 1)$-distribution

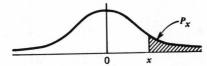

$$P_x = \Pr\{N(0, 1)\text{-variable} \geq x\}$$

P_x	x	P_x	x	P_x	x	P_x	x	P_x	x
.50	0.00	**.050**	**1.64**	.030	1.88	.020	2.05	**.010**	**2.33**
.45	0.13	.048	1.66	.029	1.90	.019	2.07	.009	2.37
.40	0.25	.046	1.68	.028	1.91	.018	2.10	.008	2.41
.35	0.39	.044	1.71	.027	1.93	.017	2.12	.007	2.46
.30	0.52	.042	1.73	.026	1.94	.016	2.14	.006	2.51
.25	0.67	.040	1.75	**.025**	**1.96**	.015	2.17	**.005**	**2.58**
.20	0.84	.038	1.77	.024	1.98	.014	2.20	.004	2.65
.15	1.04	.036	1.80	.023	2.00	.013	2.23	.003	2.75
.10	**1.28**	.034	1.83	.022	2.01	.012	2.26	.002	2.88
.05	**1.64**	.032	1.85	.021	2.03	.011	2.29	**.001**	**3.09**
								0.000	∞

Bold-face values are those often (but not exclusively) used when testing hypotheses and/or establishing confidence intervals.

Source: Table 2 of Lindley and Miller (1958), *Cambridge Elementary Statistical Tables*, published by Cambridge University Press, with kind permission of the authors and publishers.

The diagram, and this presentation of the table, are taken from Table 1 of S. R. Searle (1971) *Linear Models,* published by John Wiley and Sons.

Table A.2. Values of $t_{n,\alpha}$ on the $t(n)$-distribution

$t(n)$-distribution

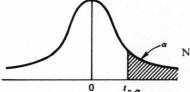

Note: $\Pr\{t(n)\text{-variable} \geq t_{n,\alpha}\} = \alpha$
$\Pr\{t(n)\text{-variable} \leq -t_{n,\alpha}\} = \alpha$
$\Pr\{|t(n)\text{-variable}| \geq t_{n,\alpha}\} = 2\alpha$

n (d.f.)	α				
	.10	.05	.025	.010	.005
1	3.08	6.31	12.71	31.82	63.66
2	1.89	2.92	4.30	6.97	9.92
3	1.64	2.35	3.18	4.54	5.84
4	1.53	2.13	2.78	3.75	4.60
5	1.48	2.02	2.57	3.36	4.03
6	1.44	1.94	2.45	3.14	3.71
7	1.42	1.89	2.36	3.00	3.50
8	1.40	1.86	2.31	2.90	3.36
9	1.38	1.83	2.26	2.82	3.25
10	1.37	1.81	2.23	2.76	3.17
12	1.36	1.78	2.18	2.68	3.06
14	1.34	1.76	2.14	2.62	2.98
16	1.34	1.75	2.12	2.58	2.92
18	1.33	1.73	2.10	2.55	2.88
20	1.32	1.72	2.09	2.53	2.84
30	1.31	1.70	2.04	2.46	2.75
40	1.30	1.68	2.02	2.42	2.70
60	1.30	1.67	2.00	2.39	2.66
120	1.29	1.66	1.98	2.36	2.62
∞ [$N(0, 1)$]	1.28	1.64	1.96	2.33	2.58

Source: Table 2 is adapted from Table III of Fisher and Yates (1963), *Statistical Tables for Biological, Agricultural and Medical Research*, 6th Ed., published by Oliver and Boyd, Edinburgh, with kind permission of the authors and publishers.

The diagram, and this presentation of the table, are taken from Table 2 of S. R. Searle (1971) *Linear Models,* published by John Wiley and Sons.

Table A.3a. Values of $F_{n_1,n_2,\alpha}$ on the $F(n_1, n_2)$-distribution

$F(n_1, n_2)$-distribution

$\Pr\{F(n_1, n_2)\text{-variable} \geq F_{n_1,n_2,\alpha}\} = \alpha = .05$

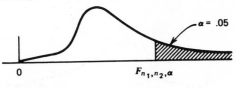

$\alpha = .05$

n_2 (denom. d.f.)	n_1(numerator d.f.)								
	1	2	4	6	8	10	12	24	∞
	$[t_{n_2,.025}]^2$				Values of $F_{n_1,n_2,\alpha}$				
1	161.4	199.5	224.6	234.0	238.9	241.9	243.9	249.1	254.3
2	18.51	19.00	19.25	19.33	19.37	19.40	19.41	19.45	19.50
3	10.13	9.55	9.12	8.94	8.85	8.79	8.74	8.64	8.53
4	7.71	6.94	6.39	6.16	6.04	5.96	5.91	5.77	5.63
5	6.61	5.79	5.19	4.95	4.82	4.74	4.68	4.53	4.36
6	5.99	5.14	4.53	4.28	4.15	4.06	4.00	3.84	3.67
7	5.59	4.74	4.12	3.87	3.73	3.64	3.57	3.41	3.23
8	5.32	4.46	3.84	3.58	3.44	3.35	3.28	3.12	2.93
9	5.12	4.26	3.63	3.37	3.23	3.14	3.07	2.90	2.71
10	4.96	4.10	3.48	3.22	3.07	2.98	2.91	2.74	2.54
11	4.84	3.98	3.36	3.09	2.95	2.85	2.79	2.61	2.40
12	4.75	3.89	3.26	3.00	2.85	2.75	2.69	2.51	2.30
13	4.67	3.81	3.18	2.92	2.77	2.67	2.60	2.42	2.21
14	4.60	3.74	3.11	2.85	2.70	2.60	2.53	2.35	2.13
15	4.54	3.68	3.06	2.79	2.64	2.54	2.48	2.29	2.07
20	4.35	3.49	2.87	2.60	2.45	2.35	2.28	2.08	1.84
25	4.24	3.39	2.76	2.49	2.34	2.24	2.16	1.96	1.71
30	4.17	3.32	2.69	2.42	2.27	2.16	2.09	1.89	1.62
40	4.08	3.23	2.61	2.34	2.18	2.08	2.00	1.79	1.51
60	4.00	3.15	2.53	2.25	2.10	1.99	1.92	1.70	1.39
120	3.92	3.07	2.45	2.17	2.02	1.91	1.83	1.61	1.25
∞	3.84	3.00	2.37	2.10	1.94	1.83	1.75	1.52	1.00

Source: Abridged from Table 18 of Pearson and Hartley (1954), *Biometrika Tables for Statisticians, Volume I,* published at the Cambridge University Press for *Biometrika* Trustees, with kind permission of the authors and publishers.

The diagram, and this presentation of the table, are taken from Table 4a of S. R. Searle (1971) *Linear Models,* published by John Wiley and Sons.

Table A.3b.   Values of $F_{n_1, n_2, \alpha}$ on the $F(n_1, n_2)$-distribution

$F(n_1, n_2)$-distribution

$\Pr\{F(n_1, n_2)\text{-variable} \geq F_{n_1, n_2, \alpha}\} = \alpha = .01$

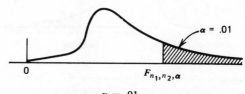

$\alpha = .01$

n_2 (denom. d.f.)	n_1 (numerator d.f.)								
	1	2	4	6	8	10	12	24	∞
	$[t_{n_2, .005}]^2$				Values of $F_{n_1, n_2, \alpha}$				
1	4052	5000	5625	5859	5982	6056	6106	6235	6366
2	98.50	99.00	99.25	99.33	99.37	99.40	99.42	99.46	99.50
3	34.12	30.82	28.71	27.91	27.49	27.23	27.05	26.60	26.13
4	21.20	18.00	15.98	15.21	14.80	14.55	14.37	13.93	13.46
5	16.26	13.27	11.39	10.67	10.29	10.05	9.89	9.47	9.02
6	13.75	10.92	9.15	8.47	8.10	7.87	7.72	7.31	6.88
7	12.25	9.55	7.85	7.19	6.84	6.62	6.47	6.07	5.65
8	11.26	8.65	7.01	6.37	6.03	5.81	5.67	5.28	4.86
9	10.56	8.02	6.42	5.80	5.47	5.26	5.11	4.73	4.31
10	10.04	7.56	5.99	5.39	5.06	4.85	4.71	4.33	3.91
11	9.65	7.21	5.67	5.07	4.74	4.54	4.40	4.02	3.60
12	9.33	6.93	5.41	4.82	4.50	4.30	4.16	3.78	3.36
13	9.07	6.70	5.21	4.62	4.30	4.10	3.96	3.59	3.17
14	8.86	6.51	5.04	4.46	4.14	3.94	3.80	3.43	3.00
15	8.68	6.36	4.89	4.32	4.00	3.80	3.67	3.29	2.87
20	8.10	5.85	4.43	3.87	3.56	3.37	3.23	2.86	2.42
25	7.77	5.57	4.18	3.63	3.32	3.13	2.99	2.62	2.17
30	7.56	5.39	4.02	3.47	3.17	2.98	2.84	2.47	2.01
40	7.31	5.18	3.83	3.29	2.99	2.80	2.66	2.29	1.80
60	7.08	4.98	3.65	3.12	2.82	2.63	2.50	2.12	1.60
120	6.85	4.79	3.48	2.96	2.66	2.47	2.34	1.95	1.38
∞	6.63	4.61	3.32	2.80	2.51	2.32	2.18	1.79	1.00

Source: Abridged from Table 18 of Pearson and Hartley (1954), *Biometrika Tables for Statisticians, Volume I*, published at the Cambridge University Press for the *Biometrika* Trustees, with kind permission of the authors and publishers.

The diagram, and this presentation of the table, are taken from Table 4b of S. R. Searle (1971) *Linear Models,* published by John Wiley and Sons.

Table A.4a. The distribution of Durbin-Watson d
5 percent significance points of d_L and d_U

	k = 1		k = 2		k = 3		k = 4		k = 5	
n	d_L	d_U	d_L	d_U	d_L	d_U	d_L	d_U	d_L	d_U
15	1.08	1.36	0.95	1.54	0.82	1.75	0.69	1.97	0.56	2.21
16	1.10	1.37	0.98	1.54	0.86	1.73	0.74	1.93	0.62	2.15
17	1.13	1.38	1.02	1.54	0.90	1.71	0.78	1.90	0.67	2.10
18	1.16	1.39	1.05	1.53	0.93	1.69	0.82	1.87	0.71	2.06
19	1.18	1.40	1.08	1.53	0.97	1.68	0.86	1.85	0.75	2.02
20	1.20	1.41	1.10	1.54	1.00	1.68	0.90	1.83	0.79	1.99
21	1.22	1.42	1.13	1.54	1.03	1.67	0.93	1.81	0.83	1.96
22	1.24	1.43	1.15	1.54	1.05	1.66	0.96	1.80	0.86	1.94
23	1.26	1.44	1.17	1.54	1.08	1.66	0.99	1.79	0.90	1.92
24	1.27	1.45	1.19	1.55	1.10	1.66	1.01	1.78	0.93	1.90
25	1.29	1.45	1.21	1.55	1.12	1.66	1.04	1.77	0.95	1.89
26	1.30	1.46	1.22	1.55	1.14	1.65	1.06	1.76	0.98	1.88
27	1.32	1.47	1.24	1.56	1.16	1.65	1.08	1.76	1.01	1.86
28	1.33	1.48	1.26	1.56	1.18	1.65	1.10	1.75	1.03	1.85
29	1.34	1.48	1.27	1.56	1.20	1.65	1.12	1.74	1.05	1.84
30	1.35	1.49	1.28	1.57	1.21	1.65	1.14	1.74	1.07	1.83
31	1.36	1.50	1.30	1.57	1.23	1.65	1.16	1.74	1.09	1.83
32	1.37	1.50	1.31	1.57	1.24	1.65	1.18	1.73	1.11	1.82
33	1.38	1.51	1.32	1.58	1.26	1.65	1.19	1.73	1.13	1.81
34	1.39	1.51	1.33	1.58	1.27	1.65	1.21	1.73	1.15	1.81
35	1.40	1.52	1.34	1.58	1.28	1.65	1.22	1.73	1.16	1.80
36	1.41	1.52	1.35	1.59	1.29	1.65	1.24	1.73	1.18	1.80
37	1.42	1.53	1.36	1.59	1.31	1.66	1.25	1.72	1.19	1.80
38	1.43	1.54	1.37	1.59	1.32	1.66	1.26	1.72	1.21	1.79
39	1.43	1.54	1.38	1.60	1.33	1.66	1.27	1.72	1.22	1.79
40	1.44	1.54	1.39	1.60	1.34	1.66	1.29	1.72	1.23	1.79
45	1.48	1.57	1.43	1.62	1.38	1.67	1.34	1.72	1.29	1.78
50	1.50	1.59	1.46	1.63	1.42	1.67	1.38	1.72	1.34	1.77
55	1.53	1.60	1.49	1.64	1.45	1.68	1.41	1.72	1.38	1.77
60	1.55	1.62	1.51	1.65	1.48	1.69	1.44	1.73	1.41	1.77
65	1.57	1.63	1.54	1.66	1.50	1.70	1.47	1.73	1.44	1.77
70	1.58	1.64	1.55	1.67	1.52	1.70	1.49	1.74	1.46	1.77
75	1.60	1.65	1.57	1.68	1.54	1.71	1.51	1.74	1.49	1.77
80	1.61	1.66	1.59	1.69	1.56	1.72	1.53	1.74	1.51	1.77
85	1.62	1.67	1.60	1.70	1.57	1.72	1.55	1.75	1.52	1.77
90	1.63	1.68	1.61	1.70	1.59	1.73	1.57	1.75	1.54	1.78
95	1.64	1.69	1.62	1.71	1.60	1.73	1.58	1.75	1.56	1.78
100	1.65	1.69	1.63	1.72	1.61	1.74	1.59	1.76	1.57	1.78

Table A.4b. The distribution of Durbin-Watson d
1 percent significance points of d_L and d_U

	$k = 1$		$k = 2$		$k = 3$		$k = 4$		$k = 5$	
n	d_L	d_U	d_L	d_U	d_L	d_U	d_L	d_U	d_L	d_U
15	0.81	1.07	0.70	1.25	0.59	1.46	0.49	1.70	0.39	1.96
16	0.84	1.09	0.74	1.25	0.63	1.44	0.53	1.66	0.44	1.90
17	0.87	1.10	0.77	1.25	0.67	1.43	0.57	1.63	0.48	1.85
18	0.90	1.12	0.80	1.26	0.71	1.42	0.61	1.60	0.52	1.80
19	0.93	1.13	0.83	1.26	0.74	1.41	0.65	1.58	0.56	1.77
20	0.95	1.15	0.86	1.27	0.77	1.41	0.68	1.57	0.60	1.74
21	0.97	1.16	0.89	1.27	0.80	1.41	0.72	1.55	0.63	1.71
22	1.00	1.17	0.91	1.28	0.83	1.40	0.75	1.54	0.66	1.69
23	1.02	1.19	0.94	1.29	0.86	1.40	0.77	1.53	0.70	1.67
24	1.04	1.20	0.96	1.30	0.88	1.41	0.80	1.53	0.72	1.66
25	1.05	1.21	0.98	1.30	0.90	1.41	0.83	1.52	0.75	1.65
26	1.07	1.22	1.00	1.31	0.93	1.41	0.85	1.52	0.78	1.64
27	1.09	1.23	1.02	1.32	0.95	1.41	0.88	1.51	0.81	1.63
28	1.10	1.24	1.04	1.32	0.97	1.41	0.90	1.51	0.83	1.62
29	1.12	1.25	1.05	1.33	0.99	1.42	0.92	1.51	0.85	1.61
30	1.13	1.26	1.07	1.34	1.01	1.42	0.94	1.51	0.88	1.61
31	1.15	1.27	1.08	1.34	1.02	1.42	0.96	1.51	0.90	1.60
32	1.16	1.28	1.10	1.35	1.04	1.43	0.98	1.51	0.92	1.60
33	1.17	1.29	1.11	1.36	1.05	1.43	1.00	1.51	0.94	1.59
34	1.18	1.30	1.13	1.36	1.07	1.43	1.01	1.51	0.95	1.59
35	1.19	1.31	1.14	1.37	1.08	1.44	1.03	1.51	0.97	1.59
36	1.21	1.32	1.15	1.38	1.10	1.44	1.04	1.51	0.99	1.59
37	1.22	1.32	1.16	1.38	1.11	1.45	1.06	1.51	1.00	1.59
38	1.23	1.33	1.18	1.39	1.12	1.45	1.07	1.52	1.02	1.58
39	1.24	1.34	1.19	1.39	1.14	1.45	1.09	1.52	1.03	1.58
40	1.25	1.34	1.20	1.40	1.15	1.46	1.10	1.52	1.05	1.58
45	1.29	1.38	1.24	1.42	1.20	1.48	1.16	1.53	1.11	1.58
50	1.32	1.40	1.28	1.45	1.24	1.49	1.20	1.54	1.16	1.59
55	1.36	1.43	1.32	1.47	1.28	1.51	1.25	1.55	1.21	1.59
60	1.38	1.45	1.35	1.48	1.32	1.52	1.28	1.56	1.25	1.60
65	1.41	1.47	1.38	1.50	1.35	1.53	1.31	1.57	1.28	1.61
70	1.43	1.49	1.40	1.52	1.37	1.55	1.34	1.58	1.31	1.61
75	1.45	1.50	1.42	1.53	1.39	1.56	1.37	1.59	1.34	1.62
80	1.47	1.52	1.44	1.54	1.42	1.57	1.39	1.60	1.36	1.62
85	1.48	1.53	1.46	1.55	1.43	1.58	1.41	1.60	1.39	1.63
90	1.50	1.54	1.47	1.56	1.45	1.59	1.43	1.61	1.41	1.64
95	1.51	1.55	1.49	1.57	1.47	1.60	1.45	1.62	1.42	1.64
100	1.52	1.56	1.50	1.58	1.48	1.60	1.46	1.63	1.44	1.65

Index

Adjusted multiple correlation coefficient, 198
Advertising data, 157, 163, 167-172
Airlines injury data, 40
Air pollution study, 209-213
Aitchison, J., 33, 50
Anscombe, F. J., 8, 9, 18, 27, 50, 107, 122
Autocorrelation, 123-142
  due to missing variables, 131-136
  effects on analysis, 123-124
  methods of detection, 125-128
  removal by transformation, 128-131
  test for, 127

Backward elimination procedure, 201-205
Bacteria deaths data, 32
Barone, J. L., 2, 18
Bartlett, M. S., 38, 50
Beaton, A. E., 2, 18
Berkson, J. A., 120, 121, 122
Best set of variables, 196
Beta coefficients, 170
Bias parameter, ridge regression, 182, 185-186, 190-192
Biased estimation, 175-192
Binomial distribution, 38
Binomial variate, 116
Bioassay, 115
Box, G. E. P., 141, 142
Brown, J. A. C., 33, 50
Brownlee, K. A., 18, 70

Characteristic roots, 162, 168, 173-174, 188
Characteristic vectors, 162, 168, 173
Chi square distribution, 54, 72
Civil Rights Acts of 1964, 144

Cochran, W. G., 18, 59, 70
Cochrane, D., 128, 129, 131, 142
Cochrane-Orcutt procedure, 128-129
Coefficients, least squares estimates, 52, 72
Coleman, J. S., 144, 172
College expense data, 103
Collinear data, 173-174
  variable selection, 206-214
Collinearity, 199-200
Confidence interval, 5, 54
  β_0, β_1, 5
  predicted values, 7, 55
Constrained estimation, 68
Constrained regression parameters, 68
Consumer expenditure data, 124-132
Correlation coefficient, 6, 55
Covariance of b_i and b_j, 54
Cox, D. R., 117, 122
Criteria for evaluating equations, 197-199
Cross section data, 85
C_p statistic, 198-199
  definition, 198
  plot, 199
Cuttoff test score, 92
Cutoff values, 86

Daniel, C., 9, 18, 50, 108, 122, 199, 214
Data sets, advertising, 158
  airlines injuries, 40
  bacteria deaths, 32
  education expenditure, 96, 108
  EEO, 146-147
  housing starts, 133
  import, 152
  industrial supervisors, 44
  money stock, 124
  pre-employment testing, 86

repair times, 10
  additional data, 15
salary survey, 75
ski equipment sales, 138
supervisor performance, 59
taxicity of Rotenone, 120
television ratings, 20
Decay rate, 32
Deficiency of data, 144, 163
Deletion of variables, effect on prediction,
  195-196, 217
Dempster, A. P., 188, 192
Dependent variable, 2, 51
Designed experiment, 117
Dose response curve, 115
Draper, N. R., 9, 18, 100, 106, 122, 214
Dummy variables, 74, 95, 137, 139
Durbin-Watson statistic, 125-128, 131, 133-
  134, 136-137
  limitations, 136-139
  tables, 222-223

Education expenditure data, 96, 108
EEO data, 145-146, 156, 162
Eigenvalues, 162
Ellenberg, J., 27, 50
Equal opportunity in education, 144
Estimated regression coefficient, 3, 52
Estimate of variance, 4, 53
European Common Market, 151, 154
Explanatory variables, 2, 51
  assumptions, 58
  measurement error, 58
  nonstochastic, 58
Exponential curve, 31
Exponential function, 31

F-distribution, 57
  table, 220-221
Finney, D. J., 118, 121, 122
Fitted value, 53
Foreward selection procedure, 201-204
F-ratio, 57
French import data, 176-179
Friedman, M., 124
Full model, 56, 215
Functional specification, 193
Furnival, G. M., 200, 214

Gallant, A. R., 31, 50

Goldberger, A. S., 38, 50
Goodness of fit, 6
  weighted least squares, 112

Heteroscedastic, 38, 101, 102, 118
  errors, detection, 45
  removal, 47
Hildreth, C., 130
Hocking, R., 188, 195, 197, 200, 214
Hoerl, A. E., 188, 214
Homoscedasticity, 38, 40
Housing starts data, 132-136

Import data, 151-152, 156, 161, 163-165,
  183-187
Incorrect model specification, effects of,
  215
Independent variable, 1, 2, 51
Index of fit, 6, 55
Indicator variable, 74, 87, 137
Industrial establishments data, 44
Interaction effects, 78
Interaction terms, 85
Iterative estimation, autocorrelated errors,
  129-131

Johnston, J., 18, 124, 128, 142

Kendall, M. G., 38, 50, 160, 172, 175, 188
Kennard, R. W., 188, 214
Kerlinger, F. N., 95, 100
Kmenta, J., 18, 95, 123, 130, 142

Lack of fit, 106
La Motte, L. R., 200, 214
Latent roots, 162
Least squares, 3, 52, 71
  estimators, properties, 53, 71-73
Linear functions of regression coefficients,
  precision, 169-170, 174
Linear model, 3, 27, 28
Logarithmic transformation, 32, 35
Logistics response function, 117, 118
Logistic transformation, 121
Logit analysis, 117
Lognormal distribution, 33, 50
Longley, J. W., 2, 18
Lu, J. Y., 130

McCallum, B. T., 175, 188

McDonald, G. C., 209, 214
Malinvaud, E., 151, 152, 172
Median dose, 117, 119
Meiselman, D., 124
Misspecification, 163, 194-196, 215-217
Model, 2
full, 56
multiple regression, 51
nonlinear, 28
reduced, 56
Money stock data, 124-132
Mosteller, F., 144, 172
Moynihan, D., 144, 172
Multicollinearity, 64, 76, 143-170
corrections for, 163-166
detection, 155-157
effects in forecasting, 151-155
effects in inference, 144-151
and ridge regression, 182-185
Multiple correlation coefficient, 55, 64, 198
Multiple regression, 1, 51-73
Multiple regression model, 51
Multiplicative random error, 33

Nonadditive effects, 78
Nonlinear model, 28
Normal equations, 52, 72, 181
Normal probability plot, 9

Observational study, 117
Orcutt, G. H., 128, 129, 131, 142
Orthogonal set, 208-209, 213
Orthogonal variables, 143, 161, 177
Outliers, detection and deletion, 19, 25

Pedhauzer, E. J., 95, 100
Pharmacology, 115
Pierce, D. A., 141, 142
Plackett, R. L., 57, 70
Poisson distribution, 38, 39, 40, 44
Predicted response, 7, 54
mean, 7, 55
values, 54
Pre-employment test data, 86
Press, S. J., 160, 172
Principal components, 157-163, 167-174, 190-192
computations, 170-172
matrix development, 172-174

multicollinearity, 157-163, 172
transformation weights, 162, 168
Principal components regression, 175-181
interpretation, 177-180
Probit model, 118
Proportion of explained variation, 6, 55, 56
Pseudoreplications, 107
Pure error, 106

Qualitative variables, 74-85

R^2, 6, 55-56
Random disturbance, 3
Rao, C. R., 18, 57, 70
Reduced model, 56
Regional differences, 109
Regression, multiple, 51-73
simple linear, 1-19
Regression coefficients, 1
partial, 51
Regression equation, 1
uses, 2, 196-197
Regression parameters, 3, 53
stability over time, 96
Repair times data, 10, 15
Replicated measurements, 105
Residual, analysis of, 9
observed, 4, 53
standardized, 4
Residual mean square, 197-198
Response variable, 1, 2, 51
Ridge estimates, 181-182, 187, 189
Ridge regression, 175, 181-192, 208-214
coefficients, 181-182, 187, 189
in matrix notation, 188-192
Ridge trace, 181-186, 190, 192, 210, 212
Rosner, B., 27, 50
Rubin, D. B., 2, 18

Salary survey data, 75
Schatzoff, M., 188, 192
Scheffe, H., 94, 100
Schwing, R. C., 209, 214
Searle, S. R., 18, 57, 70, 94, 100
Seasonal effects, dummy variables, 139-142
Seasonality, 95, 139-140

Seber, G. A. F., 57, 70
Selection of variables, 69, 193-217
Shrinkage estimators, 190-192
Silvey, S. D., 166, 172, 174
Ski equipment sales data, 137-142
Smith, H., 9, 18, 106, 122, 214
Snedecor, G. W., 18
Specification error, 143
Standardized regression coefficient, 178
Standardized variable, 134, 136, 168,
    170-171
Stepwise method, 202-203
Stepwise procedures, 201-203
Stuart, A., 38, 50
Student's t distribution, 5, 54
    table, 219
Supervisor performance data, 59, 203-
    206
Systems of regression equations, 85

Television rating data, 20
Tests of hypothesis, 5, 56
    β_1, β_0, 5
    constraints on coefficients, 68
    equality of regression coefficients, 66
    subset of regression coefficients, 65
Time series data, 85

Transformation, 35, 40
    logarithmic, 32, 35
    square root, 40
Transformation of data, 19
Transformations, 29, 38
    linearizable curves, 29
    variance stabilization, 29, 38-39
Tukey, J. W., 27, 50
Two stage estimation, 105, 110

Variable deletion, 194-196, 215-217
Variable selection, 193-217
    all possible equations, 200
    C_p statistic, 205-206
    rationale, 195
    ridge trace, 208-214
    stepwise methods, 201-203
    stopping rules, 203-205
Variance covariance matrix of b, 72
Variance inflation factor, 182, 183, 186
Variance of regression coefficients, 4, 53

Weighted least squares, 49, 101-115
Wermuth, N., 188, 192
Wilson, R. W., 200, 214
Wood, F. S., 9, 18, 50, 108, 122, 199,
    214